A Middle Way

A Middle Way

*A Non-Fundamental Approach
to Many-Body Physics*

Robert W. Batterman

OXFORD
UNIVERSITY PRESS

Oxford University Press is a department of the University of Oxford.
It furthers the University's objective of excellence in research, scholarship,
and education by publishing worldwide. Oxford is a registered trade mark of
Oxford University Press in the UK and certain other countries.

Published in the United States of America by Oxford University Press
198 Madison Avenue, New York, NY 10016, United States of America.

© Oxford University Press 2021

Cataloging-in-Publication data is on file at Library of Congress

ISBN 978–0–19–756861–3

DOI: 10.1093/oso/9780197568613.001.0001

1 3 5 7 9 8 6 4 2

Printed by Integrated Books International, United States of America

For the dogs

for the dogs

Acknowledgments

I have been thinking about these ideas for a number of years now and a lot of people have been kind enough to listen to my musings.

First off I would like to thank Leo Kadanoff, posthumously, for encouragement, support, and especially for his inspirational contributions to the subjects that I find fascinating. I feel very lucky to have gotten to know him a little since the publication of my first book. Likewise, I have benefited greatly from many interactions with Nigel Goldenfeld over the last sixteen years or so.

I owe a large debt to my friend and colleague Mark Wilson. We have been talking about philosophy, science, and mathematics for the past twenty-eight years. He has had a major influence on the way I think about problems in the intersection of these areas.

A number of my students, both past and present, provided detailed and constructive comments. I am most grateful to Julia Bursten, Kathleen Creel, Michael Miller, and Travis McKenna for their valuable contributions to this project. My co-author on a number of papers related to issues discussed in this book, Sara Green, gave me some of the most helpful critiques both in the early stages of writing and at the end. This book has been considerably improved thanks to her advice.

Several colleagues were also very helpful, for which I am most grateful. Erica Shumener helped me understand some aspects of the recent metaphysical conceptions of fundamentality. Porter Williams gave valuable criticism and, along with

Mike Miller, helped me see the importance for this project of Julian Schwinger's work on an "engineering approach" to particle physics. James Woodward offered much by way of encouragement, support, and valuable critiques. Colin Allen read drafts of all of the chapters and gave me excellent feedback. Likewise, I am honored that my friend Roger Jones took the time to read the work and give exceptionally constructive criticism.

I owe a great deal to Laura Ruetsche and Gordon Belot for clubbing me several times as the project progressed. They pushed me very hard to clarify various aspects of the arguments and to dial back some of my more outlandish suggestions. For this I am deeply appreciative.

I would also like to thank several anonymous readers and my editor from Oxford University Press, Peter Ohlin, for encouraging me over the years.

Finally, I thank Devi and Quinn for keeping me sane; but not Carolyn, who did not.

Preface

This book focuses on a method for exploring, explaining, and understanding the behavior of many-body systems. These are large systems consisting of many components that display distinct behaviors at different scales. They include gases, fluids, and composite materials such as wood and steel. In the context of condensed-matter physics this method was described in a famous paper by Leo P. Kadanoff and Paul C. Martin entitled "Hydrodynamic Equations and Correlation Functions." It describes an approach to non-equilibrium behavior that focuses on structures (represented by correlation functions) that characterize mesoscale properties of the systems. In other words, rather than a fully bottom-up approach, starting with the components at the atomic or molecular scale, the "hydrodynamic approach" aims to describe and account for continuum behaviors by largely ignoring details at the "fundamental" level.

This methodological approach actually has its origins in Einstein's work on Brownian motion. There Einstein made two pioneering arguments. First, he gave what may be the first instance of up-scaling or homogenization to determine an effective (continuum) value for a material parameter—the viscosity—by considering the heterogeneous mixture of the solvent and the Brownian particles. It turns out that this method is of a kind with much work in the science of materials. This connection and the wide-ranging interdisciplinary nature of these methods are stressed.

Einstein's second argument led to the first expression of a fundamental theorem of statistical mechanics called the Fluctuation–Dissipation theorem. This theorem, whose importance was only really appreciated much later, provides the primary justification for the hydrodynamic/correlation function methodology.

The hydrodynamic methods exploit the fact that there *must* be heterogeneous structures at mesoscales in between the atomic and the continuum. Such structures can be employed to describe and explain the behaviors of many-body systems in near but non-equilibrium states. In particular, the methodology focuses on understanding transport behavior— currents that appear as a result of spatial and temporal non-uniformities with respect to conserved quantities. This allows for an understanding of non-equilibrium behavior that is not based on bottom-up derivations starting with the Boltzmann equation.

Furthermore, the hydrodynamic methodology allows for a novel explanation of the relative autonomy of upper-scale continuum theories like fluid mechanics, from lower-scale, more fundamental theories. We can understand the remarkable fact that such continuum theories survived the atomic revolution despite being ontologically inaccurate. These theories treat systems as having no structure below the continuum scales, but remain spectacularly successful in their engineering applications to real-world problems.

Reflections on this methodology led me to an argument for treating the correlational mesoscale structures as natural kinds. The argument is neither based on philosophical intuitions about the nature of laws and lawfulness, nor on metaphysical considerations of fundamentality and joint carving. I argue that there are scientific reasons to treat the parameters characterizing the relevant aspects of the mesoscale as the right or natural variables for studying heterogeneous many-body systems

I argue that the Fluctuation-Dissipation theorem along with the other considerations in the book may be seen yielding metaphysical consequences. The arguments support a claim that mesoscale structures, and parameters that describe them, can be/should be considered, in a certain sense, more fundamental than the lowest scale atomic details. I provide an argument to the effect that the mesoscale parameters and variables are the *right* variables for doing condensed matter physics broadly construed. They are *natural*. In metaphysical terms, they better carve nature at its joints than lower scale, presumably more fundamental structures.

Overall, this book argues for a middle way between continuum theories and atomic theories. Except for very special cases, the reductionist goal of providing direct connections between atomic theory and continuum mechanics and thermodynamics is bound to fail. A proper understanding of those inter-theory relations can be had when mesoscales are taken seriously.

There are places in this book where the material becomes somewhat technical. This is particularly the case in Chapter 3 on correlation functions and the hydrodynamic description of many-body systems. I have included the equations and derivations for readers who wish to follow the detailed steps of the arguments. But, I hope that the philosophical implications of the hydrodynamic methodology are sufficiently salient that readers who wish to skim over these parts of the book may do so without loss of understanding.

Most importantly, I hope this book demonstrates the importance of the mesoscale, middle-out, approach to physical systems, and that it offers a new philosophical perspective on understanding behaviors of systems across scales.

Contents

Chapter 1

Introduction

Philosophers of science, particularly philosophers of physics, have focused their attention on *foundational problems* in physical theories. They have also examined the relations between *non-fundamental* theories and those considered to be *fundamental* or more *fundamental* than others. The terms "foundational" and "fundamental" get bandied about, used sometimes interchangeably, and sometimes in different ways. As a rough characterization, I shall take "fundamental" primarily to refer to theories reflecting some kind of hierarchy with a (the) fundamental theory being the bottom or perhaps the "ground" upon which higher, less fundamental, theories may depend.[1] I shall take "foundational" to refer to types of deep problems that lie at the core of an individual theory or at the interface between a pair of theories. These problems are most often of a conceptual or broadly logical nature. (More about this in the next section.) Metaphysicians of science want, to some extent, to discuss the fundamental. Philosophers of specific physical theories often write about foundational problems or puzzles in those theories.

---

[1] The existence of dependence relations (reductive relations, for example) and the existence of some one fundamental theory (a theory of everything) are, of course, deeply disputed claims. I just note that fact for now.

A Middle Way: A Non-Fundamental Approach to Many-Body Physics. Robert W. Batterman, Oxford University Press. © Oxford University Press 2021.
DOI: 10.1093/oso/9780197568613.003.0001

On the assumption that the distinction between "foundational problems" and "fundamental vs. non-fundamental theories" is reasonably cogent, this book primarily addresses the latter. I believe that the focus on foundational problems in physical theories has led many philosophers of physics (myself, at times, included) to fail to appreciate some important aspects of physical methodology that have been developed in the study of many-body systems construed most broadly. This is primarily because attention to foundational problems of a conceptual nature directs our gaze toward the proper logical structure of the theory and its proper axiomatic formulation, rather than toward the processes by which scientists actually develop models and theories. I believe that this is residue from the days of logical positivism and logical empiricism.

The study of many-body systems is much more than the study of the correct axiomatic structure of classical theories or quantum theories. Some of the most successful and profound methods for gaining understanding of collective properties of many-body systems treat such systems from a *field theoretic* perspective. This perspective focuses on structures at scales intermediate between the atomic scale and the continuum scale of fluid mechanics and thermodynamics. Following condensed-matter theorists, we can refer to the mesoscale descriptions as "hydrodynamic descriptions." My hope is that by focusing on these methods we can gain considerable insight into relations between fundamental and non-fundamental theories. Specifically, I will argue that mesoscale quantities and parameters are, for the purpose of understanding the bulk behavior of many-body systems, much superior to quantities and parameters at fundamental atomic scales. They are, I hope to argue, the *natural* variables or parameters for that purpose.

1.1 Philosophy and Foundational Problems

Philosophy of physics has in practice largely concerned itself with foundational problems in physical theories. Examples of foundational problems from statistical mechanics and thermodynamics include the problem of irreversibility and the justification of appeals to ergodicity in the context of equilibrium statistical mechanics. The first concerns how to account for the manifest temporal asymmetries in the world at the scale of everyday objects, given the complete temporal *symmetry* of the dynamics governing the behaviors of the atomic and molecular constituents of such everyday objects.[2] The second involves the justification of identifying time averages with ensemble averages in various statistical calculations of equilibrium properties.[3] Quantum mechanics provides many such intractable foundational puzzles: What is the proper way to understand measurement in the theory? What is the nature of the (in)famous non-locality in that theory? Is the theory really incompatible with the existence of hidden variables? Relativistic theories also raise foundational problems concerning the causal structure of spacetime. Of current (and deep interest) are the connections between general relativistic theories of gravity and quantum field theories. There are extreme difficulties in reconciling these two theories at high energies or short distances (Burgess, 2004; Donoghue, 2012).

As noted, these problems often hint at potential "logical flaws" at the very core of scientific theories. My conception of "foundational problems" depends partly upon the presence of such logical issues.[4] Philosophers and mathematical physicists tend to approach some of these problems by trying to firm up the logical foundations of the theories. Sometimes these attempts involve trying to find the proper axiomatization

---

[2]See (Sklar, 1993).

[3]See (Moore, 2015; Malament and Zabell, 1980; Earman and Rédei, 1996; Batterman, 1998; Sklar, 1993; Werndl and Frigg, 2015a,b).

[4]I am being purposively vague about the nature of these logical flaws and issues. (Hence, the use of scare quotes.) The idea is that there are conceptual conflicts that appear hidden until philosophical spotlights are focused on core tenets of the theories. I believe the focus on logical inconsistencies has its roots in the positivist and empiricist approaches to science developed in the early part of the twentieth century. An historical analysis into this is beyond the scope of this investigation.

for the theory—one that eliminates any inconsistencies or incompatibilities. Attempts like these connect the conception of a *foundational* problem to the proper articulation or formulation of the/a relevant *fundamental* theory.

The axiomatic/algebraic program in quantum field theory provides a very nice example of this connection. That approach aims to find the proper axioms so as to place quantum field theory on firm logical grounds[5] (Halvorson and Mueger, 2006). I think it is fair to say that a large part of philosophy of physics has been directed toward investigating real and presumed "logical" inconsistencies in various theories. Additionally, philosophers of physics have aimed to find necessary and sufficient conditions that justify appeals to certain properties (e.g. ergodicity, locality, etc.). I also think it is fair to say that considerable progress has been made on many of these foundational problems.

It is also clear, as the example of axiomatic quantum field theory demonstrates, that at least one route to addressing *foundational* problems is to properly axiomatize the *fundamental* theory. In this sense, there is a relatively direct connection between attempting to solve foundational problems and Hilbert's 6th problem of axiomatizing physical theories (specifically mechanical theories). Here is Hilbert:

> The investigations on the foundations of geometry suggest the problem: *To treat in the same manner, by means of axioms, those physical sciences in which mathematics plays an important part; in the first rank are the theory of probabilities and mechanics.* (Hilbert, 1902, p. 454)

Hilbert notes that Boltzmann's work *Lectures on the Principles of Mechanics* (*Vorlesungen über die Principe der Mechanik*) "suggests the problem of developing mathemat-

---

[5]In so doing, one aim is to avoid the "inconsistencies" accused of leading to divergences and unwanted infinities in perturbative quantum field theories.

ically the limiting processes, there merely indicated, which lead
from the atomistic view to the laws of motion of continuua."

Hilbert posed a second axiomatization problem:

> Conversely one might try to derive the laws of the
> motion of rigid bodies by a limiting process from
> a system of axioms depending upon the idea of
> continuously varying conditions of a màterial fill-
> ing all space continuously, these conditions being
> defined by parameters. (Hilbert, 1902, p. 454)

This problem has received little attention by philosophers.
However, the homogenization methods investigated in this
book, while not starting from axioms, do attempt to show
how material behaviors of continuum bodies can arise from
limiting considerations starting with composites of different
materials exhibiting diverse properties. Furthermore, these
different properties are defined in terms of parameters, as we
shall see.

Consider the problem of connecting the atomistic theories
to the continuum theories via axiomatization and limiting theorems. If the right axiomatization can be found along with the
proper limit theorems, we would be able to derive *directly* the
laws of continuum mechanics from the atomistic fundamental
theory. Presumably, having done this we would also have an
understanding of how (for example) irreversible continuum-
scale behaviors are grounded in the time-symmetric laws of
the atomic theory. I believe it is fair to say that this project
has not been completely successful.

On the other hand, we will see in the course of this
book that there are genuine reasons to hold that part of
the connection of lower-scale to continuum-scale behaviors
can be realized by a focus on intermediate (meso-) scales. In
fact, I believe that philosophical discussions concerning inter-
theory and inter-level relations have focused on a limited set of
possibilities (Nagel, 1961; Batterman, 2002; Dizadji-Bahamani
et al., 2010; Schaffner, 2013; Fodor, 1974; Kim, 1992). For
example, there are Nagel-like type-type reductions where a

non-fundamental theory is presumably reduced to (derived from) a more fundamental one. There are weaker proposals where a non-fundamental theory is said to supervene on a more fundamental one, presupposing token-token reductions. In general though, discussions of inter-theory/inter-level relations have primarily sought *direct relations* between the reducing, more fundamental theory and the reduced, less fundamental theory.

I intend to argue that these supposed direct relations are rarely possible. I also intend to present a field-theoretically motivated methodology—a *hydrodynamic methodology*—for making *indirect*, mesoscale mediated connections between continuum theories and more fundamental atomic and molecular theories. This methodology is widespread and guides work in physics, materials science, and even biology. Furthermore, it works despite the fact that it almost completely ignores fundamental atomic and molecular details. It may be understood, as noted, to be a partial realization of Hilbert's converse axiomatization problem.

1.2 Autonomy and Fundamentality

There is a feature of many theoretical hierarchies that stand in the relation "theory X is more fundamental than theory Y" that standard reductive strategies generally ignore or have missed altogether. This is the fact that many less fundamental theories are to certain extents *autonomous* from their more fundamental partners. The reductive strategies ignore this autonomy, as their main goal is to show how the upper-scale, less fundamental theories, are *derivable from, and hence dependent upon* their more fundamental counterparts.

As an example of the kind of autonomy to which I refer, consider the relation between fluid mechanics and molecular dynamics: There is a deep sense in which fluid mechanics—a theory that treats fluids as continuous blobs with no structure at all—is autonomous from more fundamental theories that recognize that fluids are actually composed of various kinds of molecules. The ontologically incorrect theory of continuous blobs works amazingly well, despite the fact that fluids are actually composed of collections of discrete molecules. This is a fact, a datum, that we want to understand and explain. The general structure of this explanation is discussed in Chapter 2. Much of this book is devoted to various strategies and methodologies that play a role in realizing this understanding.

There is an irony here when one compares the previous talk of fundamentality and this notion of autonomy with some contemporary theorizing in the metaphysics of fundamentality. It seems that many metaphysical discussions connect fundamentality with some notion of (in)dependence. Thus, Tahko, noting that there is a host of (in)dependence relations D_i, formulates various conceptions of fundamentality based upon the following intuition:

> A common way to think about fundamentality is in terms of independence, whereby for any notion of dependence D, an entity is D-fundamental if and only if it does not depend$_D$ on anything else (or on anything else that does not depend on it).
>
> (Tahko, 2018, p. 3)

Of course, we are focusing on theories here, but given that the theories are descriptive of the entities, *prima facie*, we might try to order our hierarchy of theories using (in)dependence relations according to the previous schema. If we do this, as a means for elucidating the relation "theory X is more fundamental than theory Y," we run directly into the following situation. Many of our scientific theories (continuum theories like thermodynamics and fluid mechanics) are considered to be phenomenological and therefore non-fundamental. Aims to demonstrate this are the focus of the various proposals for theory reduction mentioned previously. But, as noted, these non-fundamental theories often exhibit a kind of relative autonomy or *independence* from the lower-scale theories that are often considered to be more fundamental. Thus, this kind of autonomy or relative/conditional *in*dependence[6] does not (at least not easily) map onto some metaphysicians' discussions of fundamentality in terms of independence. Upper-scale phenomena of the sort we are considering often display a remarkable insensitivity to changes in lower-scale details.

In fact, if we pay attention to actual methods from condensed-matter physics and from materials science, we will see that connections between more fundamental and less fundamental theories are mediated by intermediate mesoscale or mesolevel theories and models. Often, one does not need (nor can one obtain) many details from the fundamental theory to make upscaling connections. This is a very good thing as much of physics would not be possible if all the details were required. This is not just a matter of practical convenience, as ignoring lower-scale details and focusing on intermediate mesoscales (as much of condensed-matter physics does) actually allows us to understand the autonomy just mentioned. Furthermore, such a focus will suggest that a better account of what makes one theory more fundamental than another can be had by giving up on direct reductive connections in favor of indirect orderings based on relations of relative autonomy. These latter connections give us a different way of understanding what makes one theory more fundamental than another. It is neither a reductive relation nor a metaphysical relation of fundamentality in terms of independence.

1.3   Two-ish Senses of Fundamental

It is worth noting that Clifford Truesdell[7] makes a distinction between "structure theories" and "continuum theories" that

---

[6]See the discussion of Woodward on "the problem of variable choice" in Chapter 7.

[7]Truesdell was an idiosyncratic, sometimes irascible mathematician, natural philosopher, and physicist. He made great strides in developing continuum mechanics and putting it on reasonably firm foundations. He

also maps onto a distinction, respectively, between fundamental and less fundamental theories (Truesdell and Noll, 1992, pp. 5–8). We can get a sense of the distinction by focusing on his scorn for "real physicists"—those who exclusively study structure theories:

> The training of professional physicists today puts them under heavy disadvantage when it comes to understanding physical phenomena, much as did a training in theology some centuries ago. Ignorance commonly vents itself in expressions of contempt. Thus "physics", by definition, is become exclusively the study of the structure of matter, while anyone who considers physical phenomena on a supermolecular scale is kicked aside as not being a "real" physicist. Often "real" physicists let it be known that all gross phenomena easily could be described and predicted perfectly well by structural theories; that aside from the lack of "fundamental" (i.e. structural) interest in all things concerning ordinary materials such as water, air, and wood, the blocks to a truly "physical" (i.e. structural) treatment are "only mathematical".
>
> (Truesdell, 1984, p. 47)

The idea that the fundamental structural theories are *in principle* able to describe and predict all gross (continuum) properties is, of course, a hallmark of reductionist thinking. I believe it reflects much of the motivation for talking about fundamental vs. non-fundamental theories in the first place. Over the course of this book, we will find a number of reasons to be skeptical about this *in principle* claim.

Still another important (though related) sense of "fundamental" reflects the fact that some theories and some methods are epistemically more trustworthy than others. They lead to

---

was the founder and editor of the journals *Archive for Rational Mechanics and Analysis* and the *Archive for History of Exact Sciences*.

true, universalizable, and empirically verifiable claims about the world. It is often assumed, I believe, that the most fundamental theory in the first sense will automatically be the one that is fundamental in its trustworthiness.[8] One lesson to be learned from the discussions to follow is that this assumption is almost always mistaken.

It is often the case that deriving predictable, observable consequences from fundamental theories is, if not impossible, then extremely difficult. One can dismiss this difficulty by appeal to "in principle" derivability as just noted, but such dismissals are rarely, if ever, accompanied by a suggestion as to how this might be done. Much of science soldiers on anyway and strives to make empirically testable connections with the world. Many-body systems pose particularly difficult problems for making direct connections with observable phenomena. This is obvious, and it is clearly the reason why statistical methods in mechanics were developed by Boltzmann and Maxwell *et al.* in the first place.

As noted, I would like to focus on a particular set of methods developed for the theory of solid-state or condensed-matter physics that aim to make connections with measurement particularly direct and transparent. That is to say, these methods represent our best, most efficient, and most trustworthy means for obtaining information about many-body systems. Studying these *hydrodynamic methods* is the primary aim of this book. I believe that there are some rather deep philosophical lessons to be learned by examining the general methodology of treating many-body systems by using these techniques. Specifically, I will argue that a middle-out or mesoscale approach to studying many-body systems is superior in many contexts to methods that start with fundamental theories. In addition, I hope to show that there are genuine scientific reasons for treating mesoscale quantities or parameters as *natural* variables with which to characterize

---

[8]It is pretty clear that this assumption is one thing (there are many!) that Truesdell finds objectionable.

such systems.[9] Thus, the conclusion is that in important instances, fundamental variables are *not* the natural variables. The next section begins to lay out the ingredients required to support these claims.

1.4 Hydrodynamic Methods: A First Pass

A bottom-up approach to studying many-body systems would be to start from the fundamental description of the systems' components. In classical physics this description is provided by the complete micro-state of the system specifying the exact positions and momenta of each component, and in quantum mechanics it would be the full density matrix for the composite system. Of course, the point of view of this approach directly invites us to consider trying to solve Hilbert's sixth problem. Find the correct fundamental description/theory with its axioms and derive, via limits, the continuum description.

The hydrodynamic methods eschew this "natural" approach, adopting a point of view from (quantum) field theory. In 1959 Paul Martin and Julian Schwinger noted that from the field-theoretic point of view, large systems contain many particles, so the eigenvalue for the number operator will be very large (Martin and Schwinger, 1959, p. 1342). Most importantly they note that

> physically recognizable changes in energy are so huge compared to the energy intervals between neighboring states that the energy levels may be assumed to vary continuously. When a system is this large, the quantities of interest naturally fall into two categories. The first concerns the behavior of extensive quantities such as energy and number for which only macroscopic changes are measurable; the second refers to microscopic

---

[9]This is the subject of Chapter 7.

> features involving changes in energy and number
> that are negligible on the macroscopic scale.
>> (Martin and Schwinger, 1959, p. 1342)

This is a familiar picture of large systems whether they are classical or quantum-mechanical. Some properties or quantities are well-defined at continuum scales, and others, defined at molecular or atomic scales, are naturally unobservable at continuum scales. Thus, large many-body systems have properties that are macroscopic and microscopic and, as noted, a puzzle of continuing philosophical interest concerns how the two descriptions of such systems can be understood and related to one another.

To begin to flesh out the field-theoretic, hydrodynamic description, let us, following Kadanoff and Martin (1963), consider a large system that has been gently pushed from thermodynamic equilibrium. Equivalently, the system could find itself in this non-equilibrium state as the result of spontaneous fluctuations.[10] We assume that its physical quantities are functions of space and time and that they are slowly varying. Then the system's evolution[11]

> is completely described by local values of the various thermodynamic variables—for example, by giving the pressure, density, and velocity as a function of space and time. The basis of fluid mechanics [in this regime] is the partial differential equations satisfied by these local thermodynamic quantities.
>
> In these hydrodynamic equations, there appear a variety of parameters whose values are not given by fluid mechanics [the Navier–Stokes equations]. These parameters fall into two categories. First,

---

[10]This equivalence is extremely important. We will consider it in more detail in section 1.6.

[11]In this "hydrodynamical limit" the system is governed by five partial differential equations (Kadanoff and Martin, 1963, p. 420).

there are the thermodynamic derivatives which arise because changes in the various local variables are related by thermodynamic identities. Second, there are the transport coefficients like viscosity and thermal conductivity which enter because the fluxes of the thermodynamic quantities contain terms proportional to the gradients of the local variables. To find the values of the transport coefficients and the thermodynamic derivatives, we must turn to a *more fundamental theory* than fluid mechanics. (Kadanoff and Martin, 1963, p. 420, emphasis added)

We need to understand both what this more fundamental theory is and what this talk of "local variables" amounts to.

One way to understand the notion of local behavior is to see it as behavior in the neighborhood of some spacetime point. In the near but out of equilibrium state, one can define quantities like density around a given spatial location that differ in value compared to the equilibrium state. This allows one to define functions (correlation functions in space and time) that describe the density correlations in the neighborhood of that location. As the system evolves back to a state of equilibrium, the correlations among components will decay. Figure 1.1 illustrates that after the system (a gas in a cylinder) receives a

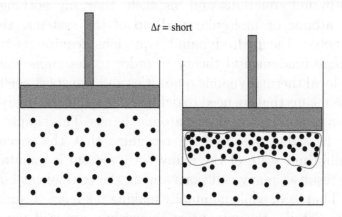

Figure 1.1: Density correlations in a pushed cylinder.

push of short duration, Δt, there is a pileup of molecules near the piston. The density of the gas there is greater than it is farther away from the piston. This pileup creates correlations among the gas molecules. Those correlations will then decay as the system returns to equilibrium with essentially uniform density of molecules throughout the (now smaller) volume of the cylinder.

In contrast, were the system near a critical point, the fluctuations would be enormous and the concept of a local neighborhood would lose its meaning. Thus, critical phenomena, and the methods by which they are understood (RG methods), are *distinct* from the hydrodynamic description we are developing here.

It is possible to formalize the concepts of local behavior and local correlations by introducing quantities called *order parameters* and *material parameters* (as we shall see later). In effect, the values for the order parameters reflect local correlations among the micro-constituents of the large system. The order parameters and the corresponding correlation functions reflect features and relations that connect upward toward the macroscopic and downward toward the microscopic. They are parameters and functions that reflect an important regime (or scale) in between the continuum and that of the atomic or molecular. They are constructions at a *mesoscale*.

The full fluid mechanical Navier–Stokes equations are continuum equations and, as such, they say nothing about the atomic or molecular makeup of the systems. However, as noted, the hydrodynamic equations require reference to a more fundamental theory in order to determine values for the local thermodynamic quantities and transport coefficients. This means that we need to define these quantities by reference to lower-scale molecular or atomic details. This simply reflects the fact that we need to recognize that the macroscopic equilibrium state of the many-body system is maintained as the result of many collisions among the component bodies.

That said, while nominally making reference to such lower-scale details, the quantities themselves are still parameters and functions that exist (or are defined) at the mesoscale mentioned previously. The local values for the thermodynamic derivatives and transport coefficients (such as viscosity) are determined through the construction and measurement of time- and space-dependent correlation functions that characterize the fluctuations away from equilibrium. They remain mesoscale functions because the information required to characterize these fluctuations is "far less specific than that contained in the microscopic state or density matrix itself" (Forster, 1990, p. 2). In other words, to describe density fluctuations one does not need to follow (or even know) the exact state of every molecule in the system. Therefore, many lower-scale details are not necessary for understanding the properties of interest (thermodynamic derivatives and transport coefficients). Things are even worse from the point of view of the "fundamental" lowest scale. Those details are also not sufficient for understanding the properties of interest: One simply cannot make sense of (nor measure) fluctuations in densities from the perspective of the lowest scale.

It is important to understand that the correlation functions are describing *patterns* in space and time. It is perhaps easiest to visualize this in terms an Ising ferromagnet. If one rapidly cools a ferromagnet from high temperatures to zero temperature, one can lock in structures—blobs of up spins and blobs of down spins. See figure 1.2. Correlation functions can describe the typical size of the blobs of up and down spins, but also, importantly their shapes and the degree to which correlations between distant spins will evolve over time. In the case of a fluid, we are similarly interested in the density correlations induced when the system is disturbed from equilibrium and in how those correlations evolve over time.

One of the most important aspects of the hydrodynamic description in terms of correlation functions is its rather direct connection with experiment. One can actually measure or probe the system to determine the nature of the correlations between spatial regions and across times. This is accomplished, for normal fluids, in scattering experiments involving light,

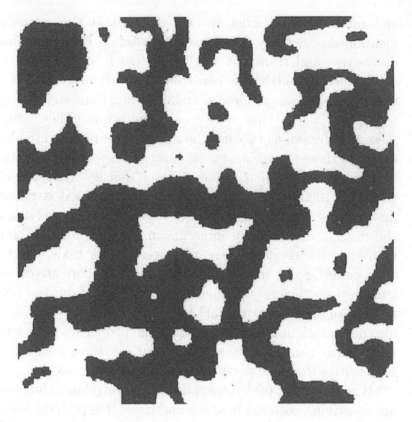

Figure 1.2: Mesoscale structures in the Ising model. Black regions spin-up; white regions spin-down. (Sethna, 2006, p. 216.)

X-rays, and neutrons. For systems with charge densities, one can scatter electrons off the material, and for magnetic systems, neutron scattering can measure the nature of correlations in the net magnetization.[12]

It is worthwhile emphasizing something rather obvious in figure 1.2 in relation to this brief discussion of measurement: The structures being probed are (as noted previously) mesoscale structures. The hydrodynamic descriptions in terms of correlation functions are clearly *not* referencing individual spins. Instead, by describing and measuring these mesoscale

---

[12]This latter example is essentially what is displayed in figure 1.2.

patterns, we are given a way to connect continuum properties like the overall magnetization with the order parameter defined at this mesoscale. We also recognize that there have to be connections between the actual distribution of up and down spins at the atomic scale and that same mesoscale order parameter. *Thus, the order parameters and the structures they describe mediate between continuum scales and atomic scales.* The nature of this "mediation" is important and will be discussed further in Chapters 4–6.

1.5 Representative Volume Elements

The recognition that in many-body systems mesoscale structures are important is an overarching theme of this book. Materials scientists have also focused on mesoscales in trying to determine continuum-scale properties of materials. A concept of great importance is the notion of a representative volume element (RVE). Let us introduce this concept by thinking about engineering structures such as a beam or a strut. Beams of different materials (steel, wood, aluminum, ...) will all exhibit different bending behaviors or will sustain different loads prior to buckling. We can experimentally determine values for various parameters that characterize the differences in their engineering capabilities. Such parameters are called "material parameters" and appear in the continuum equations used in the construction of buildings, bridges, and boats.[13] Unlike the natural philosophers of the eighteenth century, we are aware that the beams have lower-scale structure that the continuum equations completely ignore. Those natural philosophers surely suspected a connection with lower-scale details. But the proof of the existence of particulate matter came much later, largely thanks to Einstein and Perrin.

---

[13]These equations are called the Navier–Cauchy equations and are related to the Navier–Stokes equations of fluid mechanics.

Surely, it must be the case that the various values for the material parameters (such as Young's Modulus) that codify the differences in bending behaviors of wood and steel, must reflect differences in lower-scale structures between wood and steel. In other words, just as order parameters like the net magnetization code for micro-details at the mesoscale level in the context of a magnet, material parameters perform the same function in the context of materials science. How does this work?

At everyday, continuum, scales a steel beam or strut looks reasonably homogeneous. If we look at it with our naked eyes or with a magnifying glass we do not see much structure: It appears to be uniform. However, if we zoom in, using high-powered microscopes or X-ray diffraction techniques, we will begin to see structures that are hidden at everyday-length scales. In order to describe the main/dominant/important features of the steel at these shorter-length scales we employ the concept of an RVE. Consider the steel beam. A representative volume element is a material volume that is statistically representative of features of the steel at some particular spatial scale. The left part of figure 1.3 shows a "material point" surrounded by an infinitesimal material element. Structures in the material element—the voids, cracks, grain boundaries, etc.—are to be treated as the microstructure of that (macro) material element.[14]

As is evident from figure 1.3, the conception of the RVE involves the introduction of two length scales. There is the continuum or macro-scale (D) by which the neighborhood of the material point is characterized, and there is a micro-scale (d) that reflects the smallest microstructures whose properties (as we shall see, shapes are usually the most important properties) are believed to influence the overall response to stresses and strains imposed upon the neighborhood surrounding the material point. These length scales must typically differ by orders of magnitude so that ($d/D \ll 1$).

---

[14]The concept of a material point is a continuum mechanical concept and is not to be conflated with the notion of an atom or discrete element.

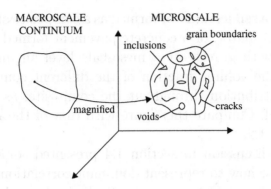

Figure 1.3: RVE. (After (Nemat-Nasser and Hori, 1999).)

It is worth flagging two features of RVEs. First, the RVE concept is *scale relative*. The actual characteristic lengths of the structures in an RVE can vary considerably. As Nemat-Nasser and Hori note, the overall properties of a mass of compacted fine powder in powder-metallurgy can have grains of micron size, so that a neighborhood of 100 microns can very well serve as an RVE. "[W]hereas in characterizing an earth dam as a continuum, with aggregates [stones, sticks, clay etc.] of many centimeters in size, the absolute dimension of an RVE would be of the order of tens of meters" (Nemat-Nasser and Hori, 1999, p. 15).

Second, the structures in the RVE *are all considered to be continua*. In other words, we are operating at scales much higher than that of the spacings of an atomic lattice, but much lower than the continuum scale at which the material appears to be homogeneous. So, actually, there are at least *three* widely separated scales. Call the scale of the atomic lattice or the scale of intermolecular interactions between fluid molecules δ_a. Then strictly speaking we have the relation:

$$\delta_a \ll d \ll D.$$

Having introduced the notion of an RVE, let us consider an example of a material that is a composite of two distinct materials. This will let us make the connection with the discussion of the previous sections much more explicit. So, consider a semiconductor formed as a composite of a conductor and an insulator, or a concrete pavement formed from mixing cement with gravel. At a mesoscale level an important feature is the volume fraction of the different components and their distribution throughout the composite, as displayed in figure 1.4. Compare this figure with that of the Ising magnet in figure 1.2.

The discussion in section 1.4 presented some theoretical details for how to represent dominant correlational structures of many-body systems using the hydrodynamic description developed by (among others) Kadanoff and Martin (1963). The similarities between the Ising model with its up spin/down spin structures and the composite depicted in figure 1.4 are striking. One should naturally suspect that similar theoretical constructs would be appropriate for characterizing the mesoscale structure of composite materials and steel or wood beams. In fact there are such methodological connections, and it is appropriate to say that the descriptions provided by RVEs in the context of materials science and in treating near-equilibrium many-body fluids and gases are largely the same. (This is discussed in some detail in Chapter 5.) The appeal to *correlation functions in space and time to describe geometric shapes and topology* are as important for condensed-matter physics as they are for materials science. These methods are

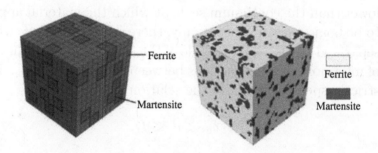

Figure 1.4: Composite RVE for steel. (Amirmaleki et al., 2016.)

ubiquitous in the study of large systems, and I want to argue that they deserve more serious philosophical consideration.

1.6   Fluctuation and Dissipation

There is one further element that is essential to the arguments to follow. We have just seen that there is a connection between the hydrodynamic methods of Kadanoff and Martin and the role played by mesoscale RVEs in materials science contexts. The connection, as noted, comes from the use and applicability of correlation functions to describe the relevant mesoscale structures in both contexts. The applicability of the hydrodynamic methods for describing non-equilibrium behavior of many-body systems is justified largely by an important theorem of statistical mechanics. This is the so-called Fluctuation–Dissipation theorem.

Many-body systems, from the perspective of statistical mechanics (as opposed to that of thermodynamics) are constantly in flux. The very notion of *equilibrium* in statistical mechanics is defined as an *average* state. Real many-body systems continually undergo spontaneous fluctuations. The Fluctuation–Dissipation theorem states an equivalence between the response of such a system to a small external disturbance (a push of some kind) and internal fluctuations in the absence of such a disturbance. Upon experiencing an external push or an internal fluctuation, the system temporarily finds itself in a non-equilibrium state. We know that it will then evolve back to equilibrium. That evolution, according to the theorem, is the same regardless of the origin of the non-equilibrium. This is quite a profound result. It says, in effect, that *equilibrium* statistical mechanics itself possesses the means for describing *non-equilibrium responses* to small external perturbations.

Fluctuations lead to correlations in quantities like the particle number that are conserved over time. For example, a fluctuation is likely to lead to spatial variations in the density of the particles that make up the system.[15] Correlation functions, as we have noted, can be used to describe such gradients in the density (and other transport properties). We expect that as the system relaxes back to equilibrium the correlations will decay and the density of the particles will again approach uniformity. The Fluctuation–Dissipation theorem asserts that the response of the system to an external push will decay in the same way. The hydrodynamic approach, as noted, allows for the determination of order parameters and material parameters that code for correlational details. In effect, the Fluctuation–Dissipation theorem guarantees the existence of such mesoscale correlational structures.

I should note here that the hydrodynamic approach to non-equilibrium behavior is distinct from *foundational* approaches based on the Boltzmann equation. The Boltzmann equation determines the temporal evolution of a density function, $f(\mathbf{r}, \mathbf{v}, t)$, of molecules in a gas.[16] It does so by taking into consideration the collisions among molecules. If $f(\mathbf{r}, \mathbf{v}, t)d\mathbf{r}d\mathbf{v}$ is the number of molecules with positions and velocities in the volume $d\mathbf{r}d\mathbf{v}$ at time t, then the Boltzmann equation takes the following form:

$$\frac{\partial}{\partial t} f(\mathbf{r}, \mathbf{v}, t) = -\mathbf{v} \cdot \frac{\partial}{\partial \mathbf{r}} f - \frac{\mathbf{F}}{m} \cdot \frac{\partial}{\partial \mathbf{v}} f + \Gamma(f) \qquad (1.1)$$

The first two terms on the right-hand side represent streaming terms that describe the free evolution of each molecule in between collisions. Here $\mathbf{v} = \dot{\mathbf{r}}$ and $\dot{\mathbf{v}} = \mathbf{F}/m$ with $\mathbf{F}$ the force on the molecule and m its mass. The $\Gamma(f)$ term is the collision term that describes the change in f due to collisions. Encoded in the collision term is Boltzmann's famous *Stosszahlansatz* which, in effect, says the the collisions are random and so the

---

[15]Consider again the piston on the right in figure 1.1. Imagine that the pileup near the piston was the result of a fluctuation away from equilibrium and not the result of a push.

[16]Here, $\mathbf{r}$ represents the spatial coordinate of the molecule and $\mathbf{v}$ represents its velocity.

system rapidly forgets correlations among molecules that the collisions naturally create.

The approach to non-equilibrium based on the Boltzmann equation has received considerable attention by philosophers (Cercignani, 1988; Sklar, 1993; Uffink, 2017). It represents a "foundational" problem in the sense introduced in section 1.1. That is to say, it is a reductive program that aims to explain the irreversibility of macro-behavior purely from the bottom up. There has been considerable progress on this foundational approach since Boltzmann's time, but it is fair to say that most successes have been for relatively sparse systems (like dilute gases) where collisions and scattering are rare, and free streaming is the dominant form of motion.

The hydrodynamic approach, in comparison, is based on the Fluctuation–Dissipation theorem. It is a mesoscale approach that privileges structures in between the atomic and the macroscopic (Kubo, 1966, p. 256). It is also an approach that easily applies to dense systems like liquids. It offers a distinct approach to non-equilibrium statistical mechanics. Unlike the bottom-up Boltzmann/kinetic theory approach, the hydrodynamic approach is restricted (to some extent) to non-equilibrium states that are near equilibrium. An advantage of the hydrodynamic approach over the kinetic theory approach is that it much more easily can provide values for the parameters and transport coefficients that characterize the upper-scale behaviors of many-body systems.

The Fluctuation–Dissipation theorem entails not only the importance, but the necessity, of mesoscale structures for understanding the relationships between continuum scale descriptions/theories and more fundamental atomic/ molecular-scale theories. The argument for this claim is the subject of Chapters 6 and 7.

1.7 Preview of Upcoming Chapters

Chapter 2 concerns the manifest relative autonomy of continuum-scale theories of large systems from lower-scale theories that reflect the actual micro-structural features of those systems. A question of interest concerns how physics can explain that autonomy. Direct upscaling approaches attempt reductive derivations of continuum properties such as density and viscosity. I argue that arguments of this stripe will fail to explain the relative autonomy and that mesoscale approaches based on renormalization and homogenization techniques are superior. The understanding of the notion of relative autonomy here is relevant to at least one important explication of what it means for one theory to be more fundamental than another.

Chapter 3 provides more detail about the hydrodynamic description of many-body systems and its equivalence to a description in terms of correlation functions. This is a dominant methodology in condensed-matter physics, and one of the core conclusions of this book is that it has very important philosophical consequences for understanding the relations of the form "theory X is more fundamental than theory Y." The discussion also elaborates on the nature of the Fluctuation–Dissipation theorem and its relation to the hydrodynamic methodology.

Chapter 4 examines the phenomenon of Brownian motion—the irregular, apparently random motion of particles suspended in a solution. In particular, it focuses on two arguments by Einstein that really reflect the core arguments of this book. In his work on Brownian motion, Einstein derived an effective viscosity for the heterogeneous system consisting of the solution plus the Brownian particles. In doing so, he essentially employed homogenization techniques for the first time. The second argument connected the dispersion of the Brownian particles (their mean square fluctuations) and the mechanism of dissipation. This was also the first instance of what later became known as the Fluctuation–Dissipation theorem. Both arguments demonstrate the absolute importance of taking a mesoscale approach to many-body systems.

Chapter 5 focuses on analogies (actually equivalences) between the upscaling arguments in the context of Brownian motion and corresponding problems in materials science and engineering. These latter include determining the effective stiffness of heterogeneous materials. Thus, the focus is on what the continuum-scale behavior of various materials will be like, given that we have some information—how much?, what kind?—concerning the nature of their lower-scale heterogeneities.

Chapters 6 and 7 argue for the primacy of a *mesoscale first* strategy for studying many-body systems. They build on work by Kadanoff, Martin, and Schwinger (among others), and develop some philosophical conclusions about methodology and about the necessity of treating order parameters and material parameters as mesoscale variables. This also connects with growing recognition, especially in biological contexts, that multi-scale systems should be modeled from a middle-out perspective. I discuss the similarities between mesoscale modeling of *inactive* materials like steel and *active* biological materials like bone. These considerations bear on how properly to understand the notion of fundamentality as discussed previously in sections 1.1–1.3.

In addition, Chapter 7 argues that the *right* variables for studying certain aspects of many-body systems are these mesoscale parameters, in that those variables are more *natural* than those that characterize the systems in terms that appear in their so-called fundamental theories. Thus, I am trying to argue that there are theoretical, scientific reasons for treating the mesoscale parameters as, in a rather strong sense, among those that should be considered natural kinds.

Chapter 2

Autonomy

In section 1.3 I discussed some ways of thinking about what makes a theory a fundamental theory. On the one hand, there is the (reductionist) view that fundamental theories are "structural" in Truesdell's sense. These are the theories of the "real" physicists that focus on the structure of matter. Such theories ignore phenomenological theories (such as continuum mechanics) that describe "less fundamental" or phenomenological behaviors of matter at everyday length and time scales. Furthermore, it is often asserted by proponents of these fundamental theories, that gross phenomena (upper-scale, everyday behaviors of many-body systems) are fully describable and predictable from the fundamental theories. Here is Truesdell expressing the reductionist position in particularly stark terms:

> [Such gross phenomena] could be described and predicted perfectly well by structural theories; that aside from the lack of "fundamental" (i.e. structural) interest in all things concerning ordinary materials such as water, air, and wood, the blocks to a truly "physical" (i.e. structural) treatment are "only mathematical". (Truesdell, 1984, p. 47)

A Middle Way: A Non-Fundamental Approach to Many-Body Physics. Robert W. Batterman, Oxford University Press. © Oxford University Press 2021.
DOI: 10.1093/oso/9780197568613.003.0002

Another sense of fundamental importance reflected the fact that some theories are epistemically more trustworthy than others. The "real" physicist quoted previously believes that the theory that is fundamental in the structural sense is also most likely to be fundamental in this second sense. After all, anything we might want to know about water, air, and wood follows mathematically (with certainty) from the fundamental structural theory.

Claims like these challenge the idea that phenomenological theories like continuum mechanics or, more generally, special sciences like psychology and economics, exhibit genuine autonomy from the more fundamental "structural" theories of high-energy physics. If all generalizations of such phenomenological theories can really be described and predicted perfectly well by structural theories, then their apparent autonomy is just that—apparent. Exactly what I mean by "autonomy" here is the subject of this chapter. For the moment it suffices to think of it as referring to the *relative* independence of *successful* upper-scale (continuum) theories of material behavior from any specification of details about the actual molecular, atomic, or subatomic structure of the materials. In other words, our theories of "ordinary materials such as water, air, and wood" work remarkably well despite completely ignoring any structure of those materials at scales below millimeters. This relative independence from lower-scale details is at the heart of the notion of autonomy under consideration in this chapter. The autonomy or independence here is not absolute. I am not claiming that the lower-scale molecular or atomic theories are completely irrelevant for the upper-scale behavior. However, the degree to which the lower-scale details do not matter is quite striking. Furthermore, as we will see, the degree to which mesoscale structures do matter to upper-scale behavior is also quite striking.

In this chapter I will elaborate on the notion of autonomy that is at play in this debate. While discussions about the relationships between "fundamental" or structural theories and non-fundamental/phenomenological theories have often been framed in terms of questions of reduction and emergence,

I believe that such framing can be misleading. It can keep us from identifying the most important issues raised by the conception of autonomy. In fact, I believe that the framework in which questions about reduction and emergence are usually raised actually hinders an investigation into what I believe is the most important feature of inter-theory relations. This is the manifest (relative) *autonomy* of the "less fundamental" theories from "more fundamental theories." Thus, the way to understanding the relation "theory X *is more fundamental than* theory Y" is through an investigation into this notion of autonomy.

To begin, let us consider an old example due to Hilary Putnam (1975) about explanatory priorities.

2.1 Pegs and Boards

Despite just saying that talk of reduction can be misleading, it is useful to develop the nature of the kind of autonomy in question by discussing a famous example of Putnam's and an influential response to the conclusions Putnam draws from that example. The response from Elliott Sober (1999) is framed in terms of inter-theoretic reduction. Putnam asks us to consider a (reasonably stiff or rigid) board containing two holes—one square of side length 1 cm, the other circular of diameter 1 cm. Next note that a square peg of side length 0.9 cm will fit through the square hole but not through the circular hole. Why? Putnam claims that the macroscopic geometric properties of the peg-and-board system explain this fact and that a deduction from the microstructure (atomic/molecular) of the board and peg will *not* explain this fact. (Putnam asserts that a long detailed quantum-mechanical description of the board and peg is completely unnecessary to explain the macroscopic behavior of this system: "If you want to, let us say that the deduction *is* an explanation, it is just a terrible explanation, and why look for terrible explanations when good ones are available?" (Putnam, 1975, p. 296).)

Sober counters that intuitions can pull one in different directions and that Putnam's claim about the explanatory priority of the macroscopic regularity is illusory.

> Perhaps the micro-details do not interest *Putnam*, but they may interest *others*, and for perfectly legitimate reasons. Explanations come with different levels of detail. When someone tells you more than you want to hear, this does not mean that what is said fails to be an explanation. There is a difference between explaining too much and not explaining at all. (Sober, 1999, p. 547)

Sober considers Putnam's claim that the macroscopic explanation is superior to the micro/quantum explanation as a challenge to a relatively standard understanding of *explanatory reductionism*. On this view the reductionist asserts:

i. "Every singular occurrence that a higher-level science can explain also can be explained by a lower-level science."

ii. "Every law in a higher-level science can be explained by laws in a lower-level science." (Sober, 1999, p. 543)

To these claims, Sober adds the following rider:

> The "can" in these claims is supposed to mean "can in principle," not "can in practice." Science is not now complete; there is a lot that the physics of the present fails to tell us about societies, minds, and living things. However, a completed physics would not be this limited, or so reductionism asserts... (Sober, 1999, p. 543)

Note that the apparent superiority of the macroscopic/geometric explanation can in part be taken to be an expression of the fact that the lower-scale quantum mechanical details are not that relevant to the observed peg-and-board behavior. Put differently, the macro-explanation is to a certain extent autonomous from any lower-scale details. One can strengthen this intuition by appealing to a multiple realizability argument made famous by Fodor (1974) in "Special Sciences, or the Disunity of Sciences as a Working Hypothesis." To do so, we need to consider two systems of pegs and boards made of different materials. (This is exactly what Sober has us consider.)

So for the sake of argument, let us assume that the first board and peg are made of a ferrous material, like iron, and that the second system is made of some non-ferrous material, such as aluminum. These differences might very well affect the behavior of the pegs as they go through the square holes, as there may be magnetic effects in the iron peg-and-board system absent in the aluminum system. If we adopt Putnam's macro-explanation, then we will have the same explanation for the pegs' behavior in the two cases. This has the advantage of providing a "unified" explanation of the different systems' behaviors. On the other hand, if we opt for a micro-explanation, then, since the pegs and boards are different, the micro-details and hence the micro-explanations will likewise be different. In such a case we will have a less unified or a "disunified" explanation (Sober, 1999, pp. 550–551). Is the choice between providing a unified or a disunified explanation of the pegs' behavior an *objective* choice between two genuinely competing explanations? Sober says "no."

> ... I am claiming that there is no objective reason to prefer the unified over the disunified explanation. Science has room for both lumpers and splitters. Some people may not be interested in hearing that the two systems are in fact different; the fact that they have the same macro-properties may be all they wish to learn. But this does not show that discerning differences is less explanatory. Indeed, many scientists would find it more illuminating to be shown how the same effect is reached by different causal pathways. (Sober, 1999, p. 551)

We see that Sober again counters by claiming that the choice between the unifying and disunifying explanation is a pragmatic choice.

The multiple realizability argument challenges the reductionist claims expressed in (i) and (ii) by highlighting the *autonomy* of the higher-level (geometric) explanation from the lower-level (*very different*) micro-based explanations stemming from the microscopic details of the iron and aluminum boards and pegs. As we see, Sober thinks this autonomy is, at best, only apparent, since the higher-level singular occurrences and the higher-level generalizations are reductively explainable by lower-level physics.[1] I think that this misses the real point of the challenge of multiple realizability.[2]

I believe that the real challenge posed by the possibility of multiple realizability is to provide an answer the following question:

- (**AUT**)[3] How can systems that are heterogeneous at some (typically) micro-scale exhibit the same pattern of behavior at the macro-scale?

Note that this question refers to a macroscopic *pattern* of behavior. This is important. Patterns are repeatable and relatively robust phenomena. So (**AUT**) is asking for an account of a repeatable, relatively stable phenomenon. As

---

[1]Sober's and Fodor's takeaways from the multiple realizability argument are completely at odds. Sober thinks the argument against reductionism fails and that at least, *in principle*, one can show that the upper-level physics is not autonomous from lower-level theory. Fodor, to the contrary, believes it shows the *absolute* autonomy of the upper-level physics from the lower-level physics.

[2]This is not to say that Fodor, Putnam, or Sober actually addressed the challenge I am posing. In fact, I think they have not, and that is one reason why the debates about multiple realization and reduction still continue in the literature.

[3]In "Autonomy of Theories: An Explanatory Problem" (2018), upon which some of this chapter is based, I called this question "**MR**." As it is at heart a question about explaining the relative autonomy of upper-scale behavior from lower-scale detail, I think the current designation is more appropriate.

such, the question concerns the very possibility of this stability under variations in lower-scale detail.

Now we can ask the following: Do the "disunified" explanations actually provide an answer to this question? For that matter, does the "unified" explanation actually provide an answer to this question? I contend that neither do. And so, Sober and those, like Jeremy Butterfield (Butterfield, 2011a,b) who endorse his argument, have overlooked an important explanatory challenge.

Consider the two micro-explanations of the pegs' behavior relative to the boards'. The first peg, call it "A," needs to be described in all of its quantum-mechanical glory. Since (for now) we are considering explanations to be "in principle" explanations, we can assume at this point that such a description can indeed be provided.[4] Next, A's state description serves as input in the appropriate dynamical equation (the Schrödinger equation) from which we are to imagine we can derive its trajectory through the square hole in the first board. Similarly, a different state description of the second peg, "B," serves as input for determining the behavior of B's trajectory through the square hole in the second board. Of course, we are going to need extreme micro-descriptions (quantum descriptions) of the two boards as well. Given the differences in materials (iron vs. aluminum), these descriptions will, likewise, be very dissimilar. All this is just to say that the macro-behavior of the two systems is multiply realized by heterogeneous realizers.

These distinct derivations are completely disjoint. The derivation of A's behavior tells us nothing about the behavior of B, and *vice versa*. In what sense do we provide an explanation for the common macro-scale behavior of these two peg-and-board systems by performing these *in principle* derivations? One might respond by claiming that providing both derivations together explains the common behavior of the two systems. But this misses the key aspect of the question

---

[4]Let me be clear that I believe that ultimately this assumption lacks sufficient support. More about this in section 2.3.1.

(**AUT**); namely, that we are interested in *understanding what is responsible for the robustness or stability* of the macro-scale pattern of behavior. The disjoint explanations, even taken together, do not address this question.

I suggest that the only way to answer this is to provide an account of why the *details* that genuinely distinguish these systems from one another at smaller scales, are irrelevant for the macroscopic behavior of interest.[5] In doing this one is able to demonstrate that the macroscopic behavior is stable under changes in the microscopic details. And, with the demonstration of that stability, one can understand how it is that the macro-behavior is relatively autonomous from the micro-details. Neither of these derivations provide anything like such an account.

Does the upper-level unified explanation provide an answer to our question? Here too I think that the answer is "no." An appeal to geometric properties together with the rigidity of the pegs and boards does explain why peg A can proceed through the square hole and not through the round hole. Similarly, for the behavior of peg B. Does this explain *how multiple realizability is possible or why* (**AUT***) is true* according to the theory that distinguishes the realizers? No. Rather, it describes the behavior to be explained in non-fundamental terms. It appeals to the fact that the diagonal of the peg is greater than the diameter of the round hole. If we are interested in why pegs and boards exhibit this exclusionary behavior *despite the fact that they have different microstructures*, we do not have an answer in terms of the reducing theory alone.

The challenge of multiple realizability to explanatory reductionism *properly understood,* concerns the ability of the theory of the heterogeneous micro-realizers to explain the *robustness* of the common behavior displayed by the systems at macro-scales. That is, the challenge is to explain the *autonomy* of upper-scale common behavior from lower-scale

---

[5]For example, these details tell us that the microstructure of iron and aluminum are relevant for determining their distinctive magnetic properties.

details. However, as we have seen, "disunified" explanations, while certainly telling us a lot about the behavior of individual systems, do not explain the autonomy in question. And this is true even if we buy into the idea that some day we will have a completed physics—even if we dismiss explanatory difficulties as "merely mathematical" or as involving only "pragmatic" difficulties.

Sober's take-home message is that reductionists should

> build on the bare proposition that physics in principle can explain any singular occurrence that a higher-level science is able to explain. The level of detail in such physical explanations may be more than many would want to hear, but a genuine explanation is provided nonetheless, and it has a property that the multiple realizability argument has overlooked. For reductionists, the interesting feature of physical explanations of social, psychological, and biological phenomena is that they use the same basic theoretical machinery that is used to explain phenomena that are nonsocial, nonpsychological, and nonbiological... The special sciences unify by abstracting away from physical details; reductionism asserts that physics unifies because everything can be explained, and explained *completely*, by adverting to physical details. (Sober, 1999, p. 561)

This message, however, does nothing to answer the question (**AUT**), which, I claim, is a distinct problem concerning the explanation of the autonomy of upper-scale patterns of behavior from lower-scale detail. Neither the lumping nor the splitting strategies answer (**AUT**).

I have followed Fodor's and Sober's lead here in framing the question (**AUT**) in the context of a discussion about reduction and the special sciences. However, the key issue really concerns the fact (and it is indeed a fact) that many of our scientific theories of the behavior of systems and materials at everyday (continuum) length scales are *almost completely autonomous* of lower-scale details. This is the fact that (**AUT**) asks us to explain.

2.2 How to Answer (AUT)

Truesdell claims that "real physicists" insist the generalizations describing "gross phenomena" could easily be described and predicted by "structural"/fundamental theories. The only difficulty involved in carrying out such descriptions and predictions are mathematical. I take it that this dismissal is equivalent to the claim (expressed by Sober) that (according to explanatory reductionism) in principle completed physics would be able to explain the singular occurrences and general laws appealed to in "higher-level" sciences. The previous argument is in part meant to establish that the fact of autonomy (namely, the object of the question (**AUT**)) cannot be explained by the *standard* (derivation-of-laws-from-laws) explanations to which philosophers typically appeal. That framework, as noted previously, does not allow for an answer to questions like (**AUT**).

If the previous argument is persuasive, then there is a genuine explanatory challenge provided by the fact of multiple realizability. I believe that an answer to (**AUT**) cannot be provided by the kinds of explanatory stories presupposed by the "real physicists" with whom Truesdell finds fault. It is *not* a challenge to explanatorily reduce the higher-level science to the more fundamental "lower-level" science. Instead, *the challenge is to account for the the success of our phenomenological theories.* How can theories of continuum-scale physics (continuum mechanics, hydrodynamics, thermodynamics, etc.) work so well and be so robust when they essentially make no reference to the fundamental structures that our foundational physical theories are about?

Philosophers and physicists often conflate *in principle* explanatory reductionism with an account of the success and robustness of our phenomenological theories. I have been arguing that this is a mistake. But it does not mean that the question (**AUT**) fails to be genuine. So, rather than focus on reductive explanations, I think we should focus on answering (**AUT**).

Before getting to this, though, we should address a potential objection to the whole project.

2.2.1 Multiple Realizability? Really?

Perhaps my claim that there is a genuine explanatory challenge provided by the *fact* of multiple realizability is too hasty or even false. After all, a main argument in Polger's and Shapiro's recent *The Multiple Realization Book* (2016) is that multiple realization is hard to come by and that multiple realizability without real evidence is a philosophical cop-out:

> On our view, multiple realization should be understood as something like the claim that systems of many (even indefinitely many) different kinds of physical structures can have exactly the same psychological functions. That kind of multiple realization would be enough, at least prima facie, to raise trouble for identity theories... By the same token, if we don't find evidence of actual multiple realization, then advocates of functionalist and realization theories cannot confidently retreat to the claim that psychological states are possibly multiply realized—that they are multiply realizable. Asserting the possibility of multiple realization is not an alternative to finding evidence for that possibility. Evidence of multiple *realizability* is just as necessary as evidence of actual multiple *realization*. (Polger and Shapiro, 2016, pp. 58–59)

I do not want to weigh in on whether Polger and Shapiro are correct to characterize multiple realization in terms of *function* for the mind–brain relations they are addressing. However, I do want to argue that their account of multiple realization is too narrow to accommodate all reasonable claims about multiple realization. According to their "basic recipe," "Multiple realization occurs if and only if two (or more) systems perform the same function in different ways"[6] (Polger and Shapiro, 2016, p. 45). This rules out the possibility that there are other (non-functional) features of the world that might be multiply realized. The question (**AUT**) is a question about behaviors of systems at different scales. Specifically, it is a question about how to explain the *fact* that systems that are different at some micro-scale can display identical *patterns of behavior* at macro-scales. The question does not in any way assume that the identical patterns of behavior are performances of the same *function*. In fact, every case of multiple realization in this book is a case where the focus is on upper-scale behavior. It turns out that multiple realization is really quite ubiquitous.[7]

So, let us consider a famous example of multiple realization/realizability in physics. And let us see how physicists have tried to address the question (**AUT**) in the face of this example.

2.2.2 Universality

Physicists use the term "universality" essentially in the way philosophers use the term "multiple realizability" (Batterman, 2000, 2002). Michael Berry (1987, p. 185) has said that asserting that a property is a universal property of a system is "the slightly pretentious way in which physicists denote the identical behaviour in different systems. The most familiar example of universality from physics involves thermodynamics near critical points." Consider figure 2.1 from a famous paper by Guggenheim from 1945. This figure plots the temperature vs. density of eight different fluids in reduced (dimensionless)

---

[6]They refine this, but not in any way that affects my point here.

[7]See some of the examples in (Batterman, 2002) and the discussion in (Koskinen, 2019).

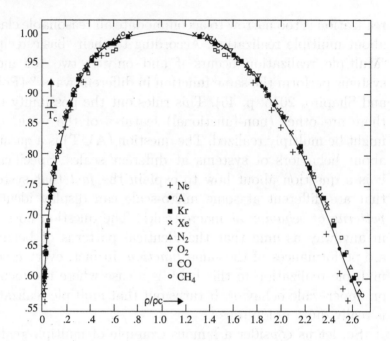

Figure 2.1: Universality of critical phenomena. (Guggenheim, 1945.)

coordinates. Values on the x-axis below 1.0 represent the density of vapor phase of the fluids, and the values above 1.0 represent the density of the liquid phases of the fluids. Thus at 1.0 the densities of the different phases are the same. The y-axis plots the (dimensionless) critical temperature of the fluids where the value 1.0 means that a system's temperature is the critical temperature. The curve in figure 2.1 is called a coexistence curve, and it provides the various densities of liquid and vapor phases at different temperatures. The remarkable thing about this plot is the fact that it shows the shape of the coexistence curve to be *the same for each fluid at its critical value for density and temperature.* Every different fluid represented has a different molecular makeup. For example, neon (Ne) and methane (CH$_4$) have very different microstructures. As a result of these different microstructures the actual critical temperatures and critical densities of each fluid will be different. Nevertheless, when one plots the behaviors of these different systems in reduced coordinates, one can see that each system exhibits identical behavior near their respective critical points—*the shape of the curve is identical for each system.* This is a paradigm example of universality/multiple realizability. Each molecularly distinct system exhibits the same macro-behavior represented by the fact that the data for each system all lie on the same curve. How is this remarkable multiply realized pattern possible? What, in other words, is the answer to (**AUT**) for this particular case?

2.2.3 Renormalization Group

It was not until the 1970s that there was a satisfactory answer to how the universal behavior of critical phenomena is possible. That answer came out of work by Leo Kadanoff, Michael Fisher, and Kenneth Wilson. Wilson won the Nobel prize for finalizing the technique that enables one to demonstrate that the (molecular) details that genuinely distinguish the different fluids from one another (that genuinely allow us to see, for example, that each has a different critical temperature and critical density) are *irrelevant* for the common macro-scale behavior of interest (that they all have coexistence curves of the same shape). This mathematical argument is called the renormalization group explanation of the universality of critical phenomena.

Let me very briefly and non-technically outline the explanatory strategy.[8] The basic physical idea is that the universal upper-scale thermodynamic behavior of systems near criticality is dominated by fluctuations in various crucial quantities, and that the range of the fluctuations is enormous in comparison to the range of the intermolecular forces governing the interactions between molecules. As a result, the fluctuations are remarkably insensitive to the detailed nature

---

[8]I have talked about this strategy in many places over the years. See (Batterman, 2000, 2002, 2011, 2019). Maybe the most useful discussion in the current context is (Batterman, 2019).

of the intermolecular forces. The renormalization group gives one a way to exploit this fact and, within certain limits, to demonstrate that differences in intermolecular forces *play essentially no role* in the upper-scale behavior. Another way to put this is that one can change the "essential molecular features" of, say, methane into those of neon, without affecting the upper-scale behavior (that is, without changing the shape of the curve in figure 2.1). This reflects a kind of *stability under changes* of the very nature of the systems. The renormalization group makes the metaphor of "morphing" one system into another mathematically precise.

It does this by first constructing an enormous abstract space, each point of which might represent a real fluid, a possible fluid, a solid, etc. Next one induces on this space a transformation that has the effect, essentially, of eliminating degrees of freedom by some kind of averaging rule. That is to say, one replaces specific interactions among a cluster of nearby molecules with averages. Then, very roughly, one considers the

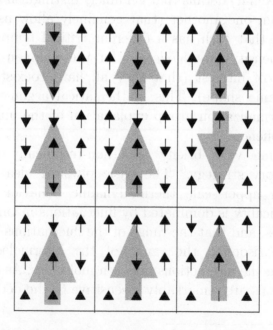

Figure 2.2: Blocking and averaging to yield a new (coarse-grained) system. (Kadanoff, 2013, p. 172.)

averages to be new "molecular components" that interact with others in a new "coarse-grained" cluster or block. Figure 2.2 gives an idea of how this works. For ease of visualization, the figure shows this for a model of spins on a lattice, but something similar will work for fluid molecules. In fact, the renormalization group argument allows us to see that this spin model (Ising model) is in the same universality class as the fluids in the Guggenheim plot! This provides a justification for our use of simple, minimal models to study behaviors of real fluids (Batterman and Rice, 2014).

The idea exploits the fact that near the critical point systems exhibit the property of self-similarity. Thus one can replace the original degrees of freedom with averages as exhibited in figure 2.2 because the self-similarity guarantees that the new system will look like the old one as one coarse-grains or zooms out. One then rescales the system in an appropriate way that takes the original system to a new *effective* system in the space of systems. This latter system still exhibits macro-scale behavior similar to the system one started with. This provides a (renormalization group) transformation on all systems in the abstract space. By repeatedly performing this operation, one eliminates more and more detail that is irrelevant for that macro-behavior. Next, one examines the topography of the induced transformation on the abstract space and searches for fixed points of the transformation. These are points which when acted upon by the transformation yield the same point.[9] A fixed point is a property of the transformation itself, and all details of the systems that flow toward that fixed point have been eliminated. Those systems/models (points in the space) that flow to the *same* fixed point are in the same universality class, and they will exhibit the same macro-behavior.[10] That macro-behavior can be determined by an

---

[9]If τ represents the transformation and p^* is a fixed point, we will have $\tau(p^*) = p^*$.

[10]To put this another way: the universality class is the basin of attraction of the fixed point.

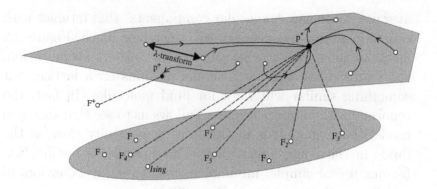

Figure 2.3: Fixed point in abstract space and universality class.

analysis of the transformation in the neighborhood of the fixed point.

In figure 2.3, the lower collection represents systems in the universality class delimited by the fact that these systems/models flowed to the same fixed point, p^*, under the appropriate (renormalization group) transformation τ in the upper abstract space. The "λ-transformation" here is the precise expression of the "morphing" talk mentioned earlier. On can effect such a transformation because the systems have been shown to be in the basin of attraction of the same fixed point. Note that another system/model, F^+ fails to flow to the fixed point p^* and so that system/model is not in the universality class.

The argument just sketched, by which one can delimit the class of heterogeneous systems all exhibiting the same macro-behavior, is not at all like the kind of derivation from initial data and fundamental equations of the kind Sober sees in the disunified explanations he discusses. Nevertheless, it is an *explanation* of how the universal/multiply realized common macro-behavior is possible from the point of view of a theory that genuinely distinguishes the realizers from one another. Note that we have neither explained a single occurrence of a higher-level property nor a higher-level law. We have provided, instead, an answer to the question (**AUT**).

2.3 Generalizations

In effect, this renormalization group argument allows one to extract continuum-scale, phenomenological behavior by systematically ignoring lower-scale details. As noted, it is particularly apt for this task in contexts where it is possible to exploit the fact that the systems exhibit a kind of self-similarity—a feature of systems near their critical points. But the question of (relative) autonomy also arises for systems that are not near criticality and where it will not be possible to exploit self-similarity. In fact, (**AUT**) is worth answering in virtually any context where we witness upper-scale continuum behavior that is robust, reproducible, and repeatable. In Chapter 1 we began to consider the hydrodynamic description of many-body systems and saw the importance of introducing quantities like order parameters that are defined at some mesoscale between the continuum and the atomic. In such contexts we can begin to upscale from the lowest levels to the highest scales via this intermediate hydrodynamic/correlation function description. These investigations also provide an answer to the question (**AUT**). They too provide a means for essentially ignoring lowest-scale details in the service of explaining upper-scale autonomy. Here I briefly describe this procedure/process from the point of view of modeling of material behaviors. We will discuss this more in Chapter 5.

2.3.1 Multi-scale Modeling of Materials

The bending behavior of a steel beam, say, is remarkably well described and explained by a continuum equation (the Navier–Cauchy equation) that was derived long before there was any empirical evidence for atoms:

$$(\lambda + \mu)\nabla(\nabla \cdot \mathbf{u}) + \rho\nabla^2\mathbf{u} + \mathbf{f} = 0 \qquad (2.1)$$

Here $\mathbf{u}$ is the displacement vector, ρ is the material density, and $\mathbf{f}$ represents the body forces (eg. gravity) acting on the material. The material parameters, λ and μ, are the "Lamé" parameters and codify the bending and stretching behaviors of the material. They take on different values for different materials.

The Navier–Cauchy equation itself makes absolutely no reference to any structure in the beam and treats it as completely homogeneous at all scales. The only empirical input to the equation comes in the form of various material parameters—the Lamé parameters (these are related to Young's modulus) and the density—that, in effect, *define* the material of interest. Values for Young's modulus are determined typically through workbench measurements of how much a material extends upon being pulled and shortens upon being squeezed. These values are clearly related somehow to the *actual* atomic and mesoscale structures (the lower-scale heterogeneity) present in the beam.

Determining the connection between these lower-scale structures and the potential values for the constitutive parameters is a difficult mathematical problem known as "homogenization."[11] At best, homogenization enables one to establish a fictitious, effective model beam that is equivalent in its bending behavior to the actual steel beam. In fact, if one can find fictitious or *effective* values for the material parameters λ and μ, then one has an explanation for why the Navier–Cauchy equations work for the steel beams at the continuum scale. There is, in essence, a fictitious homogeneous beam that will display the same behaviors at the continuum scale that the actual heterogeneous (at lower scale) beam displays. Then, from the point of view of the Navier–Cauchy equation, it is fine to treat the beam as having no lower-scale structure whatsoever.

Let me relate this to the discussion in section 2.2.3 and specifically to figure 2.3. The fictitious homogeneous system is a minimal model—one that ignores the heterogeneous details

---

[11]In fact, the mathematics involved is indeed related to the renormalization group arguments discussed in section 2.2.3. There is, as noted, no requirement of self-similarity.

of actual system. Finding such an effective homogeneous minimal model establishes a stability transformation (like the λ-transformation) that shows the behavior of the actual system to be stable under perturbation/transformation to the fictitious minimal model. Thus the homogenization methods allow for an explanation of (**AUT**) as well. In Chapter 4 we will see that Einstein, in the context of Brownian motion, provides just such an explanation.

The ideal of *in principle* derivation of behaviors of systems (or laws, or theories) from more "fundamental" lower-scale details (or more fundamental laws or theories) is largely mistaken. Any examination of the actual practice of scientists interested in modeling systems at different scales will reveal nothing as simple as the kind of derivation that proponents of this ideal believe is possible. The appeal to a completed ideal physics—the main feature that underwrites these *in principle* claims—is purely aspirational and speculative. We have no idea what such a physics would look like, nor do we have any evidence that it exists. Furthermore, the way scientists actually do attempt to answer questions of the form (**AUT**) looks nothing like the kind of *in principle* derivations to which philosophers often appeal.

Do these claims constitute an argument that a true completed physics of the sort Sober imagines in not possible? No. The RG argument and the upscaling via homogenization remain completely agnostic about the possibility of a true complete *fundamental* physics. A remarkable feature of the RG strategy is that it allows one to construct a hierarchy of effective theories even though we are completely ignorant of the physics at the highest energies (smallest spatial scales) where presumably the completed theory is to be found. There may very well be a true complete physics; but whether that is so is irrelevant for the explanatory story.

Does the lack of an argument to the effect that a true completed fundamental physics cannot exist count against the strategies for answering questions of the form (**AUT**)? I do not believe so. It seems to me that the burden of proof regarding this question should be on the individual claiming that a completed physics, if it exists, will be able to answer the explanatory question at issue. If "completed" entails being able to answers questions like (**AUT**), then that is question-begging. If "completed" does not entail this, then one would like to have some idea of how such an explanation might begin to be formulated. And I know of no place where this has been attempted.

2.4 A Brief Thought Experiment

Before concluding this chapter I would like to engage in some mildly wild speculation.

It is an empirical fact that the world we live in comes in the kind of layers that lead us to ask (**AUT**). In other words, there *is* universal/multiply realized behavior exhibited by systems that are genuinely heterogeneous at lower scales. I have been arguing that this fact requires an explanation. Reductionists will typically argue that these layers do not depend on *all* of the details at lower scales, and that averaging over lower-scale details will be sufficient to show that some kind of layer talk may be justified, although essentially eliminable. In the chapters to come, I will argue that this kind of simple averaging argument will almost always fail, and that a mesoscale first approach is required to make connections between the different layers in the actual world.

But what if the world were such that there are no layers displaying behavioral regularities that are relatively independent from lower-scale layers or, for that matter, from upper scales? What if relatively upper-scale behaviors depended essentially on all or many of the details of a more fundamental microscale or a less fundamental upper, upper scale? There is no *a priori* reason that this could not be the case. What would such a world actually be like? It would be one where there are strong correlations across scales. There would be no stable realm of phenomena for theories like thermodynamics and fluid mechanics to latch onto. The behaviors of systems in this world would depend upon details at all spatial and temporal scales. Whether or not my coffee maker would still be a coffee maker tomorrow morning would depend sensitively upon the detailed quantum state of all the atoms in Mount Rushmore.

In fact, would it even make sense to talk about systems and their behaviors? It seems there would be no real distinction between systems or behaviors at distinct scales. The very concept of behavior-at-a-scale would make no sense whatsoever. Would it even be possible to identify systems as being the same at different times and in different locations?

There are consequences for metaphysical discussions about fundamentality here, I believe. One would not be able to form hierarchies of theories fitting the schema "theory X is more fundamental than theory Y." Nor could there be hierarchies of systems or properties. If, as Tahko (2018) suggests, fundamentality is to be cashed out in terms of (in)dependence relations, in this world everything would depend upon everything else. Conceptions of fundamentality based on dependence, I contest, would not make sense.

A question to keep in mind as we progress through this book is whether there could be any *natural* properties or natural kinds in this strange world. I raise this question in part to suggest that in our world, where it *does* make sense to ask (**AUT**), the mesoscale properties upon which we will focus can be genuine candidates for natural kinds. This is the subject of Chapter 7.

2.5 Conclusion

This chapter aims to draw attention to an explanatory problem posed by the existence of multiply realized, or universal, behavior exhibited by certain physical systems. The problem is to explain how it is possible that systems radically distinct at lower scales can nevertheless exhibit identical or nearly identical behavior at upper scales (typically everyday length and time scales). I think the philosophical literature has by and large missed the fact that this is an interesting question to ask.[12] However, when philosophers consider issues relating to multiple realizability, they almost always focus on questions of reduction. Specifically, they adopt the stance taken by Clifford Truesdell's "real physicist" and by Elliott Sober (at least in (1999)). Namely, they argue that upper-scale behavior—behavior that is often characterized very nicely by the equations of continuum physics—is completely understandable and explainable in terms of *fundamental/atomic/subatomic* physics. They claim that the only obstacle to actually providing explanations of such behavior is purely "mathematical" or pragmatic. Sober, as we have seen, expresses this claim in terms of the possibility of "in principle derivability". *In principle,* "[e]very singular occurrence that a higher-level science can explain also can be explained by a lower-level science," and that *in principle,* "[e]very law in a higher-level science can be explained by laws in a lower-level science" (Sober, 1999, p. 543).

I have been arguing that these two claims are essentially vacuous. Those who make such claims need to provide at least some idea of how such derivational explanations are supposed to proceed. In addition, if one is interested in answering questions of the form (**AUT**), even if one could provide such *in principle* derivational explanations, they would not constitute an answer. To answer (**AUT**) requires a demonstration of the stability of the upper-scale (continuum) behavior under the perturbation of lower-scale (molecular/atomic/subatomic) details. Neither of these *in principle* derivations (even if possible) do this. So at the very least, if one believes that (**AUT**) is a legitimate scientific question,[13] one needs to consider different explanatory strategies. The renormalization group

---

[12]There are exceptions. Fodor (1997, pp. 160–161) asks the question and says there is no answer. Papineau (1993, p. 43) asks the question and assumes that the only response that is not "incredible" is the reductive one which I have been challenging. For an earlier response to Papineau, see (Batterman, 2000, section 7).

[13]The speculations of section 2.4 are, in part, raised to show that it is legitimate to ask (**AUT**).

and the theory of homogenization are just such strategies. They proceed by focusing on mesoscale structures in between the continuum and the atomic. They are *not* bottom-up derivational explanations.

Finally, considering the question (**AUT**) has allowed us to articulate a conception of relative autonomy of one theory/ model to another. We have fleshed out a relationship of the form "theory X is more fundamental than theory Y" that is missed by reductively motivated accounts of fundamentality hierarchies.

Chapter 3

Hydrodynamics

In the last chapter I raised the question of explaining the relative autonomy of "non-fundamental" continuum-scale physics from atomic "fundamental" physics. In the introductory chapter I suggested that a way to understand this autonomy, and autonomous relations between theories in general, starts with a mesoscale "hydrodynamic" description of many-body systems. This chapter further elaborates on some of the details of this important methodological approach. The next section introduces an example to illustrate the hydrodynamic methods. This is a relatively simple example, due to Kadanoff and Martin, of a system for which the only important conserved quantity is the total spin magnetization. I then discuss, in section 3.2, the intimate connection between the hydrodynamic description of spin transport and the correlation function description of the same system. Along the way we see the first appearance of the Fluctuation–Dissipation theorem (recall 1.6) in the guise of Onsager's regression hypothesis. Section 3.3 introduces the notion of a linear response to an external disturbance to the system.

3.1 Conserved Quantities and Transport

We are interested in many-body systems in near-equilibrium states. For example, we could consider a one component

A Middle Way: A Non-Fundamental Approach to Many-Body Physics. Robert W. Batterman,
Oxford University Press. © Oxford University Press 2021.
DOI: 10.1093/oso/9780197568613.003.0003

"normal" fluid[1] that has been pushed out of equilibrium or has spontaneously evolved into such a state by a statistical fluctuation. We would like an equation that describes the limit of slowly varying flows in such a non-equilibrium situation. The *locus classicus* providing the derivations of such (partial differential) equations is the paper by Kadanoff and Martin (1963), "Hydrodynamic Equations and Correlation Functions." I have already referred to this paper in Chapter 1, where we saw that such equations contain parameters that appear neither in the continuum equations of fluid mechanics (the Navier–Stokes equations) nor in the underlying/more fundamental equations of molecular dynamics. Instead, they play essential roles in the mesoscale/hydrodynamic theory we are considering. In this section I will first outline the generic recipe for deriving hydrodynamic equations and then develop these equations for a relatively simple specific example—spin diffusion.

The first step of the recipe is to identify quantities that are conserved under the dynamical evolution of the system. For a normal fluid like water, these quantities are the number of particles (the mass density), the momentum density, and the energy density. Let us briefly consider the mass (or number) density. In equilibrium this density will be unchanging at any point in space. But in an out-of-equilibrium situation there will be local regions in the neighborhood of a point where the density differs. Recall figure 1.1. There will then be a net flow of particles from a region of greater density to one of lesser density as the system relaxes toward its equilibrium state. We can consider this net flow or migration of particles to be a current. Given this, we can write down an equation expressing the conservation of this quantity (Kadanoff and Martin, 1963, p. 437):

$$\frac{\partial}{\partial t}n(\mathbf{r}, t) + \frac{1}{m}\nabla \cdot \mathbf{g}(\mathbf{r}, t) = 0 \qquad (3.1)$$

Here $n(\mathbf{r}, t)$ is the density of particles at a point $\mathbf{r}$ at time t, and $\mathbf{g}(\mathbf{r}, t)$ is the momentum density. Thus, $\frac{1}{m}\nabla \cdot \mathbf{g}(\mathbf{r}, t)$ is the

---

[1]That is, not a superfluid.

momentum current or the flux of the particle density. There are corresponding conservation equations for momentum and energy with corresponding currents (Kadanoff and Martin, 1963, p. 437).

Equation (3.1) by itself is not very useful. It restricts the dynamics somewhat because of the conservation of mass. However, in order to solve for the number density one needs an equation that *defines* the current in terms of the number density. Finding this definitional equation—it is called a "constitutive equation"—represents the second step in the generic recipe. With these two ingredients one can write down a partial differential equation that describes the long-wavelength, slow evolution of the near-equilibrium many-body fluid.

A simple analogy should help in understanding the concept of a constitutive equation. Suppose we write down Hooke's law for a weight on a spring:

$$F_s = -kx.$$

This is an equation that tells us that the restoring force F_s is proportional to the displacement of the spring, x, from its equilibrium at $x = 0$. As it stands, this equation can tell us nothing about the behavior of any actual spring. We need to know the value of the spring constant k which is definitive of the material. Thus, specifying k is the analog of providing a constitutive equation.

In fact, this is a rather familiar problem. Consider the conservation law for charge in electrodynamics[2] (Kadanoff, 2000, p. 79):

$$\frac{\partial}{\partial t}\rho(\mathbf{r}, t) + \nabla \cdot j(\mathbf{r}, t) = 0 \qquad (3.2)$$

where $\rho(\mathbf{r}, t)$ is the charge density and $j(\mathbf{r}, t)$ is the current density. In order to use this equation we need a constitutive

---

[2]Note that this conservation equation has the same form as equation (3.1).

equation defining the current. In electrodynamics there will be different constitutive equations depending upon the type of electrodynamic material. For a simple conductor, Ohm's law is the appropriate constitutive equation:

$$\mathbf{j} = \sigma \mathbf{E} \tag{3.3}$$

where σ is the electrical conductivity and $\mathbf{E}$ is the electric field. If the material is a plasma, then there will be a different constitutive equation.[3]

Neither Kadanoff and Martin, nor Forster (1990) who follows their presentation, start with the quite involved conservation and constitutive equations for normal fluids. Instead, they choose a much simpler example. I will follow their lead and consider the hydrodynamic description for a "fluid" of spin 1/2 particles that interact through a velocity- and spin-independent force. They note that this model system is largely realized by liquid He[3] (Kadanoff and Martin, 1963, p. 422).

3.1.1   Spin Diffusion Equations

So let us consider a large collection of particles that possess a property—spin-in-a-given-direction—such that each particle can be either spin-up or spin-down in that direction.[4] If, at a given spacetime point $(\mathbf{r}, t)$, the probability that the particle is spin-up equals the probability that it is spin-down, then the density of the net magnetization in the given direction, $M(\mathbf{r}, t)$, will be zero. On the other hand, if the probabilities differ, then there will be a value for $M(\mathbf{r}, t)$ that is proportional

---

[3]See (Kadanoff, 2000, pp. 79–80) for some details. In fact, as we shall see in our discussion of materials science (Chapter 5), constitutive equations involving material parameters (such as σ in equation (3.3)) are also required to use continuum equations like the Navier–Stokes equations or the Navier–Cauchy equations to design aqueducts and bridges.

[4]Spin is really irrelevant here. All we need are some properties like black and white that are conserved and for which at a given location the particle possesses either black or white at that point.

to that difference. In other words, we have defined $M(\mathbf{r}, t)$ as follows:

$$M(\mathbf{r}, t) = \gamma \left(\rho_+(\mathbf{r}, t) - \rho_-(\mathbf{r}, t) \right) \qquad (3.4)$$

where $\rho_\pm$ are the densities in the up/down directions at the point $(\mathbf{r}, t)$, and γ is the spin magnetic moment of each particle.

In this model, the magnetization is our conserved quantity:

$$\frac{\partial}{\partial t} \int M(\mathbf{r}, t) d\mathbf{r} = 0,$$

and the conservation equation takes the form

$$\frac{\partial}{\partial t} M(\mathbf{r}, t) + \nabla \cdot \mathbf{j}^M(\mathbf{r}, t) = 0 \qquad (3.5)$$

where $\mathbf{j}^M(\mathbf{r}, t)$ is the magnetization current. Both the magnetization (equation (3.4)) and the current are *local* quantities. They depend only upon the spin properties of particles in a neighborhood of $\mathbf{r}$ at a time t (Forster, 1990, pp. 8–9).

As previously, we need a constitutive equation that defines the current in terms of the magnetization. To think about this, ask what happens to the densities $\rho_+(\mathbf{r}, t)$ and $\rho_-(\mathbf{r}, t)$ when the system is out of equilibrium. They will both evolve toward the equilibrium state in which they are equal—in which M vanishes. So, *on average* there will be a net flow of the magnetization from regions with large values for M to regions with low values for M. This gives us our constitutive equation for the situation in which the properties of the system are slowly varying in space and time:[5]

$$\langle \mathbf{j}^M(\mathbf{r}, t) \rangle = -D\nabla \langle M(\mathbf{r}, t) \rangle \qquad (3.6)$$

Here D is a transport coefficient (the spin diffusion coefficient) for the magnetization. D is proportional to the squared

---

[5](Forster, 1990, p. 9) and (Kadanoff and Martin, 1963, p. 423).

velocity of the diffusing spins.[6] Furthermore, note that we have introduced (non-)equilibrium averages because equation (3.6) is true on average only. This contrasts with the continuity (or conservation) equation (3.5) which is exact and microscopic. Equation (3.6) is the constitutive equation that specifies the nature of the spin system. It is the analog of (3.3). (See, again, the example of Hooke's law on page 52.) Equation (3.6) and equation (3.5) together yield a diffusion equation for the magnetization:

$$\frac{\partial}{\partial t}\langle M(\mathbf{r}, t)\rangle - D\nabla^2\langle M(\mathbf{r}, t)\rangle = 0 \qquad (3.7)$$

It is important to stress again that the quantities in equations (3.6) and (3.7) are averages (hence the brackets) and, furthermore, that we are not following individual spin particles. This is, again, just to say that our description is a mesoscale description. We do not know (nor do we need to know) about the movements of any one of our spin-carrying particles. We are interested, instead, in *the imbalance and transport of spin density.*

Equation (3.7) also does not tell us anything about how the system came to be in an out of equilibrium state. A standard way to effect this change is to submit the system to an external magnetic field, h, in the direction of quantization[7] for a period of time. Mathematically, this is represented as follows:

$$h(\mathbf{r}, t) = h(\mathbf{r})e^{\epsilon t} \qquad\qquad t < 0$$
$$= 0 \qquad\qquad t > 0,$$

where ϵ is infinitesimal and positive. Prior to $t = 0$ the system is adiabatically subject to the external field, after which that field is turned off. The field induces a net magnetization

---

[6] An analogy here would be to the diffusivity constant in a heat equation for a given material.

[7] This is just the direction in which our spins are aligned or anti-aligned defined by "$\pm$" in equation (3.4).

$$\langle M(\mathbf{r}) \rangle = \chi h(\mathbf{r}) \tag{3.8}$$

with χ being the spin susceptibility[8]:

$$\chi = \left. \frac{\partial M}{\partial h} \right|_{h=0} \tag{3.9}$$

Thus, while the field is on $(t < 0)$, the magnetization obeys equation (3.8), and after it is switched off, it obeys equation (3.7).

It is relatively simple to solve the spin diffusion equation (3.7) for the bulk (systems of infinite extent). One does this respectively via Fourier transforming in space and Laplace transforming in time:

$$\langle M(\mathbf{k}, t) \rangle = \int \langle M(\mathbf{r}, t) \rangle e^{i\mathbf{k} \cdot \mathbf{r}} d\mathbf{r} \tag{3.10}$$

$$\langle M(\mathbf{k}, z) \rangle = \int_0^\infty \langle M(\mathbf{k}, t) \rangle e^{izt} dt \tag{3.11}$$

where $\mathbf{k}$ is the wave vector of the fluctuation and z is the complex frequency, which is required to be in the upper half of the complex plane for the second equation to converge. Using these transformations in equation (3.7) we can calculate $M(\mathbf{k}, z)$. Letting $h(\mathbf{k})$ be the Fourier transformed magnetic field at time t_0, the result is the following:

$$\langle M(\mathbf{k}, z) \rangle = \frac{\chi h(\mathbf{k})}{-iz + Dk^2} \tag{3.12}$$

Since the magnetization is a conserved quantity, the behavior captured by (3.12) is a spatially sinusoidal collective variable. Fluctuations in M are slow to disappear on the timescale of collisions. "A local excess of [M] cannot disappear locally (which could occur rapidly) but can only relax by spreading slowly over the entire system" (Forster, 1990, p. 11). Figure 3.1

---

[8]The susceptibility, in this context, quantifies the degree to which the system becomes magnetized in the external, applied magnetic field.

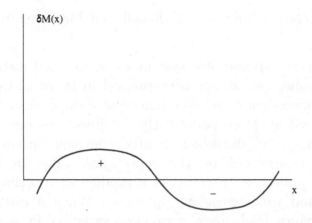

Figure 3.1: Sinusoidal spatial fluctuation in M. (After (Forster, 1990).)

shows a (sinusoidal) fluctuation in M of wavelength λ. Such a fluctuation can only relax by physically transporting the conserved quantity (the spins) from nearby regions (small values of x) to farther away regions. In terms of the wavelength of the fluctuation, this involves transport over a distance on the order of $\lambda/2$ and would take an infinite amount of time as $\lambda \to \infty$. If the transport takes the form of a random walk with the diffusion (transport) coefficient D, then the relaxation time will be on the order of λ^2/D.

The long lifetime of these fluctuations (in comparison with collision times) is a hallmark of a hydrodynamic mode. Any disturbance to a quantity that is *not* conserved will most certainly be short-lived due to the chaotic nature of the dynamics of the many-body system. As a result, such hydrodynamic modes lend themselves to probing experimentally. They are the most important patterns of behavior of many-body systems below the scale of the continuum.

3.2 Correlation Functions

One of the most important aspects of the hydrodynamic description is that it is equivalent to a description in terms of correlation functions. Kadanoff and Martin state this as follows:

> The response of a system to an external distur-
> bance can always be expressed in terms of time
> dependent correlation functions of the *undisturbed
> system*. More particularly the linear response of
> a system disturbed slightly from equilibrium is
> characterized by the expectation value *in the
> equilibrium ensemble*, of a product of two space-
> and time-dependent operators. When a distur-
> bance leads to a very slow variation in space
> and time of all physical quantities, the response
> may alternatively be described by the linearized
> hydrodynamic equations. (Kadanoff and Martin,
> 1963, p. 419, my emphases.)

Perhaps the most important fact (emphasized) expressed in this quote is that we can understand the linear response of a near equilibrium system in terms of the statistical mechanical characterization of the corresponding equilibrium system. Thus, equilibrium statistical mechanics *itself* has the means to describe the non-equilibrium behavior of transport properties in the slow, linear regime. This is why there can be an equivalence between the hydrodynamic description and that in terms of correlation functions.

This is also the core philosophical consequence of the Fluctuation–Dissipation theorem introduced in section 1.6. This theorem affirms "a general relationship between the response of a given [many-body] system to an external dis-turbance and the internal fluctuation of the system in the absence of the disturbance" (Kubo, 1966, p. 256). Put slightly differently, it means that an equilibrium fluctuation (some-thing to be expected in large systems) can tell us about how the system will respond to an external disturbance. Thus, as noted, equilibrium statistical mechanics itself has the means to tell us how a near-but-out-of-equilibrium system will behave.

In this section I examine some features of this correlation function description of many-body systems.

Let us consider our spin system as it is pushed from equilibrium (where $M(\mathbf{r}, t) = 0$) by the magnetic field $h(\mathbf{r})$ according to equation (3.8). We can plot the non-equilibrium average value for the magnetic current as a function of time. See figure 3.2. Prior to the push, $\langle \mathbf{j}^M(\mathbf{r}, t) \rangle_{n.e} = 0$; and, immediately after the field is turned on (at t), the current increases until it reaches a non-equilibrium steady state (at t').[9] Then, after the field is turned off (at t''), the average current decays back to its equilibrium value of zero. We assume that the current average, $\langle \mathbf{j}^M(\mathbf{r}, t) \rangle_{n.e}$, displays linear behavior when out of equilibrium. In fact, we assume that the non-equilibrium behavior obeys Onsager's *regression hypothesis*. According to this hypothesis the relaxation of the induced macroscopic non-equilibrium disturbance back to equilibrium follows the same laws as the regression of microscopic fluctuations at equilibrium. These microscopic fluctuations are represented by correlation functions. The Fluctuation–Dissipation theorem

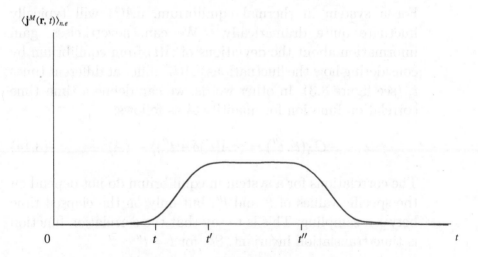

Figure 3.2: Current vs. time.

and Onsager's hypothesis establish an equivalence between the decay of the correlations and the macroscopic response of the system to an external push.

To unpack Onsager's hypothesis, consider a quantity (an observable) A that describes a property of a statistical system.[10] For example, A might represent the pressure of a gas in a box. If we think of A as a function of time (changing pressure of the gas), then in statistical mechanics we do not directly observe $A(t)$. Instead, we identify the observed macroscopic quantity with a phase or ensemble *average* taken with respect to some privileged probability distribution (the canonical distribution) appropriate for representing the system's equilibrium state. This average is time-independent. However, we can ask about the deviation of $A(t)$ from its equilibrium value, $\langle A \rangle$, with respect to the equilibrium, probability distribution:

$$\delta A(t) = A(t) - \langle A \rangle \qquad (3.13)$$

For a system in thermal equilibrium, $\delta A(t)$ will typically fluctuate quite dramatically.[11] We can, nevertheless, gain information about the deviations of $A(t)$ from equilibrium by considering how the fluctuations $\delta A(t_i)$ differ at different times t_i (see figure 3.3). In other words, we can define a time–time correlation function for quantity A as follows:

$$C_A(t', t'') = \langle \delta A(t') \delta A(t'') \rangle - \langle A \rangle^2 \qquad (3.14)$$

The correlations for a system in equilibrium do not depend on the specific values of t' and t'', but only on the elapsed time between sampling. This is to say that the correlation function is time translation invariant. So, for $t = t'' - t'$

---

[10]The following discussion is generic in that rather than continuing to treat $M(\mathbf{r}, t)$ as the observable of interest, we consider any phase function or observable of the system "A".

[11]The exception is if A is a constant of the motion governing the gas. For instance, if the gas is energetically insulated from its environment, then energy, E, will be a constant of motion and the equilibrium value for δE will be zero.

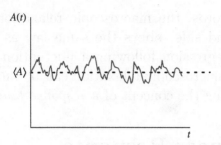

Figure 3.3: $A(t)$ vs. $\langle A \rangle$.

$$C_A(t) = \langle \delta A(t') \delta A(t'') \rangle \tag{3.15}$$

where we set $\langle A \rangle^2 = 0$ since the equilibrium average is time-independent. In the short time limit where $t = (t'' - t') \to 0$, the correlation approaches

$$C_A(0) = \langle \delta A(0) \delta A(0) \rangle = \langle (\delta A)^2 \rangle$$

At long times, as $t \to \infty$,

$$C_A(t) \to 0$$

which is just to say that $\delta A(t')$ and $\delta A(t'')$ become uncorrelated at times long on the scale of molecular collisions. This decay of correlations with time is just the *regression of microscopic fluctuations* mentioned in Onsager's hypothesis.

Now, let us denote the non-equilibrium average of $A(t)$ by $\overline{A(t)}$.[12] This is the average over initial conditions in the phase space relative to the equilibrium (canonical) measure. Unlike the equilibrium value, $\langle A \rangle$, the non-equilibrium average, like $A(t)$ itself, is a function of time. Thus, $\overline{A(t)}$ represents the *macroscopic* non-equilibrium average of A at a time t. Onsager's regression hypothesis states the following:

$$\frac{\overline{A(t)} - \langle A \rangle}{\overline{A(0)} - \langle A \rangle} = \frac{\langle \delta A(t) \delta A(0) \rangle}{\langle (\delta A(0))^2 \rangle} \tag{3.16}$$

---

[12]We reserve the brackets for equilibrium averages.

Again, in words, the macroscopic relaxation to equilibrium (the left-hand side) obeys the same law as the microscopic relaxation/regression following a fluctuation (the right-hand side). This macroscopic relaxation can be further represented by introducing the concept of a *response function*.

3.3 Linear Response

Let us recall the diffusion equation for our spin system from section 3.1.1:

$$\langle \mathbf{j}^M(\mathbf{r}, t) \rangle = -D\nabla^2 \langle M(\mathbf{r}, t) \rangle \tag{3.17}$$

We noted that to realize a near non-equilibrium state for the system we could turn on a magnetic field h which then induces a net magnetization to the system:

$$
\begin{aligned}
h(\mathbf{r}, t) &= h(\mathbf{r})e^{\epsilon t} & t < 0 \\
&= 0 & t > 0
\end{aligned}
$$

Another way to think about this would be to write down a non-equilibrium Hamiltonian that explicitly contains the time dependence of the external field. Thus, let $\mathcal{H}_0$ be the equilibrium Hamiltonian and write the non-equilibrium Hamiltonian as follows:

$$\mathcal{H} = \mathcal{H}_0 - h(t)M \tag{3.18}$$

where, as previously, M is the net magnetization and h is the applied external magnetic field.[13]

---

[13]Generically, we have the Hamiltonian

$$\mathcal{H} = \mathcal{H}_0 - f(t)A$$

with A some internal observable upon which the time-dependent external field acts and where $f(t)$ is the time dependence of that external field.

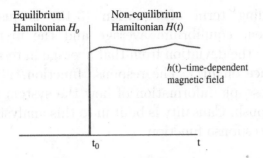

Figure 3.4: Hamiltonians with external field switched on at t_0.

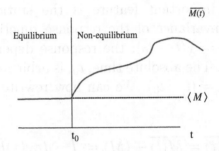

Figure 3.5: Time-dependent external field.

Figure 3.4 shows the behavior of the system Hamiltonian once the magnetic field h is turned on. (Unlike earlier, we idealize and take the switching on of the field to be immediate.) We are concerned with what happens after that instantaneous push. Figure 3.5, likewise, shows the non-equilibrium average $\overline{M(t)}$ and how it differs from the equilibrium average $\langle M \rangle$.

In the case we are considering (weak external push of the system), we can further describe $\overline{M(t)}$ by performing a Taylor expansion in powers of the perturbing field $h(t)$. Assuming that the field is applied at t_0 and that we observe the system at time $t > t_0$, this yields

$$\overline{M(t)} = \langle M \rangle + \int dt_0 \chi(t, t_0) h(t_0) + \cdots \qquad (3.19)$$

The leading term of equation (3.19) is just the time-independent equilibrium average and the first order term represents the deviation from that average in terms of a linear dependence on $h(t)$. The response function, $\chi(t, t_0)$ codifies the microscopic information of how the system reacts to the external push. Causality is built in to this analysis in the sense that the response function

$$\chi(t, t_0) = 0, \text{if} \quad t < t_0.$$

This simply means that there can be no response prior to the external push—that effects cannot precede their causes.

Another important feature is the stationarity or time translation invariance of the response function. This means that $\chi(t, t_0) = \chi(t - t_0)$; the response depends only on the time elapsed. The absolute time, t_0, is arbitrary and so we can further let $\tau = (t - t_0)$. We can now rewrite equation (3.19) as follows:

$$\delta \overline{M(t)} = \overline{M(t)} - \langle M \rangle = \int_0^\infty d\tau \chi(\tau) h(t - \tau)$$

Thus, the observable response of the system to the external push has the form of a convolution of the material response with the time evolution of the applied field.[14] Thus, the time–time correlation function is related to the response to the external push given in equation (3.8). This is just another expression of the Fluctuation–Dissipation theorem.

---

[14]This discussion is largely based on the clear presentation by Andrei Tokmakoff (Tokmakoff). I am being somewhat sloppy here using χ in this time domain representation. Actually, if we Fourier transform into the frequency domain, the transformed response function reduces to the static susceptibility equation (3.8), in the small wavenumber limit. See (Kadanoff and Martin, 1963, pp. 427–430).

3.4 Conclusion

The previous discussion is a little on the technical side. (In fact, to be completely accurate, it would actually be much more involved.) Nevertheless, I want to highlight the philosophically important aspects of the hydrodynamic description of many-body systems.

The focus is on slow responses to the system's being pushed out of equilibrium into a near-equilibrium state. The hydrodynamic description concerns so-called hydrodynamic modes (slow modes) that relate to the transport of *conserved* quantities. Non-conserved quantities (arbitrary degrees of freedom) will quickly decay after the initial push because the of chaotic nature of the microscopic interactions. Forster (1990, pp. 10–11) estimates that for a system of classical particles each of mass m governed by a pair potential interaction of strength ϵ and range a, the relaxation time τ for a non-conserved quantity is of the order

$$\tau \sim a \left(\frac{m}{\epsilon} \right)^{1/2}$$

which for Helium would give $\tau \approx 10^{-12}$ seconds. The relaxation of such degrees of freedom would be unobservable.

I started with a brief discussion of the hydrodynamic description of a system undergoing spin transport by diffusion. For that system there is one quantity that is conserved—the total magnetization. The differential form of the conservation law introduces a magnetization current seen in equation (3.5). The current is defined in terms of a diffusion coefficient.[15] By introducing a magnetic field that pushes the spin system out of equilibrium, one can provide a correlation function representation of the information contained in the spin diffusion equation. This was equation (3.12), which led us to consider the apparatus of response functions.

---

[15]This is essentially Fick's law of diffusion.

The important philosophical lessons from this discussion are the following. First, the Fluctuation–Dissipation theorem connects the response of a system to an *external* disturbance to an *internal* fluctuation of the system in the absence of any disturbance. The response to the disturbance is dissipative, as is the reaction to an equivalent internal fluctuation. Furthermore, the theorem provides the justification for the hydrodynamic approach to many-body systems. Second, the correlation function and hydrodynamic descriptions focus on quantities that describe currents or gradients of *conserved* quantities. These quantities are defined at a scale intermediate between the full fluid dynamical (Navier–Stokes) continuum theory and a molecular dynamical dynamical theory that tracks the evolution of the individual molecules that constitute the fluid. And third, although I did not discuss this except tangentially, these gradients (in magnetization density, for example) often are experimentally observable through various light, X-ray, and neutron scattering experiments. Thus, these quantities provide a rather direct connection with measurements we can actually perform on many-body systems. (We do not/cannot experimentally track the evolution of individual molecules.)

The next chapter focuses on Einstein's results on Brownian motion. This provides a concrete illustration of the importance of the Fluctuation–Dissipation theorem and of the necessity of the mesoscale, hydrodynamic approach to many-body systems.

Chapter 4

Brownian Motion

4.1 Introduction

In this chapter I consider the phenomenon of Brownian motion. This is the irregular motion carried out by particles of small size suspended in a solution. Many of us are familiar with the problem of accounting for such apparently random motion in the context of Einstein's novel theoretical reasoning that led to (among other things) a theoretical prediction for the value of Avogadro's number. This subject was, in part, the topic of Einstein's PhD dissertation and led to one of his best cited papers of the miraculous year 1905. In *Subtle is the Lord*, Abraham Pais notes that during the years 1970–1974, Einstein's Brownian motion paper was at the "top of the list of Einstein's [most heavily cited] scientific articles" (Pais, 1982, p. 89). The reason for this popularity has to do with the fact that the theoretical work is actually of much broader interest than the random motion of so-called Brownian particles. Pais continues

> ... the thesis, dealing with bulk rheological proper-
> ties of particle suspensions, contains results which
> have an extraordinarily wide range of applications.
> They are relevant to the construction industry (the
> motion of sand particles in cement mixes), to the
> dairy industry (the motion of casein micelles in

A Middle Way: A Non-Fundamental Approach to Many-Body Physics. Robert W. Batterman, Oxford University Press. © Oxford University Press 2021.
DOI: 10.1093/oso/9780197568613.003.0004

cows's milk), and to ecology (the motion of aerosol particles in clouds), to mention but a few scattered examples. (Pais, 1982, p. 90)

I note here that Einstein's results are also important to understanding various aspects of material behavior (such as the bending of beams) that, at least *prima facie*, have no obvious connections with motion in particle suspensions. This is the subject of Chapter 5.

There are two arguments that Einstein combined in his discussions of Brownian motion in the service of predicting the value of Avogadro's number. The first, discussed in the next section, addresses a problem of upscaling. Einstein showed how, given the viscosity of the solvent fluid, the addition of Brownian particles changes the viscosity in the now heterogeneous solution—i.e., the Brownian particles plus the solvent. In this context the Brownian particles occupy a mesoscale (between the continuum scale where we can talk about viscosity and the micro-scale where the "invisible particles" composing the solvent fluid reside). It is this argument that will eventually connect up with the behaviors of materials just mentioned.

This argument can also be seen to provide an answer to the question raised by (**AUT**) in Chapter 2. In other words, Einstein's derivation of the effective viscosity for the heterogeneous Brownian system demonstrates that the continuum equations (the Navier–Stokes equations) are relatively autonomous of the lower-scale details of the actual heterogeneous system. I explain how this goes in section 4.3.1.

Einstein's second argument results in a connection between the mean square fluctuations of the Brownian particle's momentum and the mechanism of dissipation that is described on the continuum scale by the effective viscosity. This latter argument is an instance of the Fluctuation–Dissipation theorem. I outline a version of this in section 4.4.

It is quite remarkable that Einstein's arguments concerning the phenomenon of Brownian motion connect the two themes of this book—the upscaling from RVEs to continuum parameters, and the justification for the mesoscale hydrody-namic/correlation function methodology for studying many-body systems.[1]

4.2   The Hydrodynamic Equation

Let us suppose a particle is moving through a gas/fluid (a many-body system). We also assume the particle is large and massive in comparison to the average molecule in the gas. Finally, we suppose that the particle is very small in compar-ison to continuum-scale lengths (such as the dimension of the box holding the gas). As a Brownian particle, it undergoes a random-looking trajectory, zig-zagging around in a jerky fashion with no apparent preferred direction. The particle experiences different types of forces from the surrounding gas. See figure 4.1. On the one hand, collisions with the gas molecules are responsible for the particle's changes in direction, and thus changes in its momentum $\mathbf{p} = m\dot{x}$. On the other hand, the particle will also experience frictional forces due to the very same surrounding ambient gas/fluid. These forces, too, will lead to momentum changes.

Consider first the collisions with the gas molecules respon-sible for the jerky momentum trajectory.[2] We can represent the changes in momentum (or the forces) as follows:

$$m\ddot{x} = \cdots + R(t) + \cdots \qquad (4.1)$$

We let $R(t)$ represents the random impulses due to many separate kicks to the particle. At this point the "$\cdots$" are

---

[1]The importance of Einstein's work on Brownian motion has also been noted in *Reductionism, Emergence and Levels of Reality: The Importance of being Borderline* (Chibbaro et al., 2014). There are strong synergies between some of my arguments here and the presentation of the material in this book.

[2]This introduction follows to some extent the presentation by Kadanoff in (Kadanoff, 2000, pp. 120–123). It is different from Einstein's own argument.

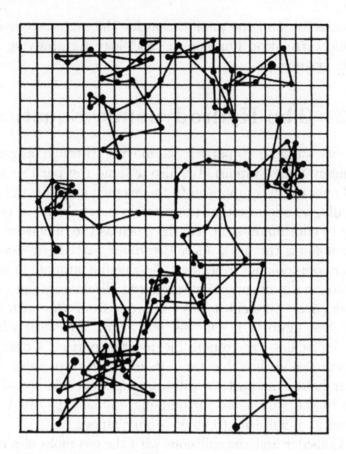

Figure 4.1: Brownian motion, Perrin's figure. (Used with permission, from (Bigg, 2008, p. 319).)

placeholders for other terms representing processes partially responsible for the momentum evolution. We assume that the mean value for R is 0, $\langle R(t) \rangle = 0$, so that there is no preferred direction of motion for the Brownian particle.[3] Another important assumption is that for any two times t_1 and t_2, there will be a correlation between the values of $R(t)$ at those times only if $|t_2 - t_1| < \epsilon$, for ϵ very small. This is a scale assumption to the effect that the molecular kicks the Brownian

---

[3]Of course, there is always gravity that would lead to an overall drift in the direction of the gravitational gradient. For our purposes here, we ignore this minor complication.

particle receives are so short in comparison to our observation time that we can register neither their temporal duration nor their nature. Thus, the contributions $R(t)$ to the momentum equation (4.1) are *uncorrelated* at the scale of observation.[4]

If the random impacts represented by $R(t)$ were the only effects on the Brownian particle by the surrounding medium, then over a long time the kicks will produce an unbounded effect—infinitely large kinetic energy will be imparted to the Brownian particle. In real systems, the particle also experiences a frictional force or drag that acts to reduce the particle's momentum. Call this force $-\eta\dot{x}$ (where η is the friction coefficient or viscosity). In the linear regime, η plays a role similar to Hooke's (linear) restoring spring constant. The hydrodynamic equation for the Brownian particle (in the absence of any external forces like gravity) will then be

$$m\ddot{x} = -\eta\dot{x} + R(t) \tag{4.2}$$

which is a simple example of a Langevin equation.[5]

We can also think about the friction term temporally. From this point of view, $-\eta\dot{x}$, is a ratio

$$\frac{\dot{x}}{\tau}$$

where τ is a relaxation time determined by the viscosity of the surrounding medium. This introduces a slow temporal scale into the momentum equation for the Brownian particle. Thinking this way, we have the schematic equation

$$m\ddot{x} = -\frac{\dot{x}}{\tau} + R(t) \tag{4.3}$$

---

[4]All of these assumptions can be relaxed. However, they were all assumed by Einstein.

[5]I should note that for short times where the Brownian particle may experience few or no collisions, the assumptions made previously will fail. In those cases we cannot assume a constant friction coefficient and would need to generalize our equations to allow for frequency-dependent viscosities. One needs to use a *generalized* Langevin equation. See (Kubo, 1966, pp. 260–263) for details.

This form explicitly demonstrates that the fast (kicking) term, $R(t)$, is partially offset by a slow (viscosity) term and the resulting evolution is one of bounded dissipation.

Kadanoff has a nice metaphor to help understand the general situation. We can consider an "inner world"—the Brownian particle—to be weakly coupled to an "outer world" of the molecules comprising the surrounding fluid which is assumed to be in statistical equilibrium.

> Let a given system not be conservative, but is instead weakly coupled to some 'outside world', which is in statistical equilibrium. Assume that there are some quantities conserved within the system itself. But our given system transfers the conserved quantities up and back to the 'outside world' via the weak coupling. In that case, if the given system is left undisturbed it eventually relaxes to a situation described by a probability distribution function which depends upon the conserved quantities, but in a very special fashion. The distribution function is an exponential function of all the exchanged extensive conserved quantities, i. e.
>
> $$\rho = \text{const.} \times exp[-\beta\mathcal{H} - \alpha N + \gamma\mathbf{v} \cdot \mathbf{G}]$$
>
> Here β, α, and γ are constants which describe the relative willingness of the system to give up energy, particles, and momentum to the outside world.[6]
>
> (Kadanoff, 2000, p. 134)

In other words, as a result of the transfer of conserved quantities back and forth between the "worlds," the system eventually reaches an equilibrium state characterized by the special probability distribution ρ.

---

[6]$\mathbf{G}$ is the total angular momentum. Both $\mathbf{v}$ and $\mathbf{G}$ are vector quantities. I have fixed some typographical errors that appear in the original.

As in the general discussion of the hydrodynamic description from Chapter 3, we see that tracking the transport of conserved quantities is crucial for understanding the behavior of the "inner world" (mesoscale) system interacting with an "outer world." The transfers of momentum, particles, etc. between worlds are exactly the structures/phenomena that the time–time correlation functions discussed in Chapter 3 are designed to describe.

The main take-away from the previous discussion is that given a Brownian particle moving in a fluid imposing frictional forces with a relaxation time τ, one can determine the average rate at which the particle accrues momentum as a result of collisions with the smaller particles constituting the fluid (Kadanoff, 2000, p. 123). However, *prima facie*, this relationship is odd. How can the increase of momentum be connected to the frictional dissipation? As we will see, that connection is just an expression of the Fluctuation–Dissipation theorem. Kadanoff notes that

> [i]t is quite natural that there is a relation between the friction and the kicks, since they are determined by precisely the same process of multiple collisions between [the] special particle and the much lighter, faster particles in the gas.
>
> (Kadanoff, 2000, p. 123)

Importantly, one sees very different effects of those very same collisions at widely separated temporal scales. This is the core aspect of the Fluctuation–Dissipation theorem and its corollary, the Onsager regression hypothesis. There is an intimate relation between the *macroscopic* (relatively long-time) disturbances away from equilibrium, and the *microscopic* (short-time) relaxation back to equilibrium resulting from spontaneous fluctuations. As noted, Einstein actually made this connection in his work on Brownian motion. This will be discussed in section 4.4.

4.3 Effective Viscosity in Brownian Contexts

Einstein's paper, "A New Determination of Molecular Dimensions" (Einstein, 1956, pp. 36–62) aims to show that

> ... the size of the molecules of a solute in an undissociated dilute solution can be found from the viscosity of the solution and of the pure solvent, and from the the rate of diffusion of the solute into the solvent, if the volume of a molecule of the solute is large compared to the volume of a molecule of the solvent.
>
> (Einstein, 1956, pp. 36–37)

Thus, his aim in this paper is to predict the size (radius) of the Brownian particle. The *solute* consists of a collection of such relatively massive particles. The *solvent* consists of the smaller molecules that constitute the ambient fluid/gas. To determine the molecular radius requires two calculations: the determination of the viscosity of the combined solute–solvent system, and a calculation of the diffusion coefficient for the solute in the solvent. The same scale assumption (large Brownian particles/small fluid molecules) is assumed here as in the last section. Einstein makes the following additional assumptions about the solute and the solvent (Pais, 1982, p. 90):

1. The combined, heterogeneous, system of the solvent plus the solute (the particles) will obey the same fluid dynamical equations that are obeyed by the solvent alone. These are the Navier–Stokes equations:

$$\rho \left(\frac{\partial \mathbf{v}}{\partial t} + \mathbf{v} \cdot \nabla \mathbf{v} \right) = -\nabla p + \eta \nabla^2 \mathbf{v} + \mathbf{f} \qquad (4.4)$$

$$\frac{\partial \rho}{\partial t} + \nabla \cdot (\rho \mathbf{v}) = 0 \qquad (4.5)$$

where η is the viscosity of the solvent, $\mathbf{f}$ are body forces such as gravity, ρ is the density of the solvent, and $\mathbf{v}$ is the velocity of the solvent.

2. One can ignore the translational inertia of the solute particles, and their rotational motion can also be ignored.

3. External forces such as gravity can be ignored (i.e., one can ignore $\mathbf{f}$ in equation (4.4)).

4. The motion of any one of the solute particles is unaffected by the motion of any other solute particle. (In other words, the solute particles are widely separated from one another.) This, too, is a spatial-scale assumption.[7]

5. The solute particles' motions are a function only of hydrodynamic stresses at their surfaces.

6. There is a no-slip boundary condition on the surfaces of the solute particles; i.e., the flow velocity $\mathbf{v}$ vanishes on the surface of the solute particles.

Einstein showed that this mixed solution can obey the fluid dynamical equations (assumption 1) only if the viscosity of the *solvent*, η, is replaced with an *effective* viscosity, η^*, for the combined solute–solvent system. This combined system is heterogeneous, and the appropriate representative volume element (RVE) reflecting that heterogeneity, once again, is at the mesoscale. He derived the form of this effective viscosity and came up with the equation:

---

[7]It also turns out that this assumption is of questionable validity. In fact, having more than one solute particle can lead to divergences in the integrals employed to calculate the effective values of the material properties like the viscosity. The problem is, that contrary to the scale separation assumption, there are long-range effects between the Brownian particles. It is possible to deal with these divergences and to properly upscale. For details, see (Jeffrey, 1977). There are, in fact, similarities between these blowups and those that appear in some perturbation expansions in QFT.

$$\eta^* = \eta(1 + \phi) \tag{4.6}$$

where ϕ is the ratio of the volume of the solute (the Brownian particles) to the volume of the solvent in some RVE large enough to be statistically representative of the mixture.[8] In other words, ϕ is the fraction of a unit volume occupied by the solute particles. Five years after Einstein published this paper, some new experimental results were achieved showing that his value for η^* was too low if equation (4.6) were correct. One of Einstein's pupils found an "elementary but nontrivial mistake in the derivation of [equation (4.6)]" (Pais, 1982, p. 92). The corrected result is the following:

$$\eta^* = \eta(1 + \frac{5}{2}\phi) \tag{4.7}$$

which led to a much improved value for Avogadro's number (Pais, 1982, p. 92).

Following Pais, we can see how, given this result, one can derive a relationship between Avogadro's number and the radius of a solute molecule. Pais notes that Einstein was able to express the volume fraction of the solute in the solvent, ϕ, as follows[9]:

$$\phi = \frac{N\rho}{m}\frac{4\pi}{3}a^3 \tag{4.8}$$

where N is Avogadro's number, a is the molecular radius, m is the molecular weight of the solute, and ρ is the amount of mass of the solute per unit volume.' (Pais, 1982, p. 90) Einstein had experimental values of $\frac{\eta^*}{\eta}$ for dilute solutions of sugar (the solute) in water (the solvent). It was also the case that the volume fraction ϕ and the molecular weight m were known. He was then able to determine a relation between Avogadro's number N and the radius a.

---

[8]Recall the discussion in Chapter 1, section 1.5.

[9]Pais' notation is in certain instances different from that of Einstein's. For example, the molecular radius in Einstein's paper is represented by P; here it is a.

A second relation between N and a had already been determined by Einstein in his 1905 paper "On the Movement of Small Particles Suspended in a Stationary Liquid Demanded by the Molecular-Kinetic Theory of Heat" (Einstein, 1956, pp. 1–18). There he had derived a relation between the diffusion coefficient for the solute molecules in a solvent and the viscosity of the solvent fluid.[10] Combing these two equations with two unknowns, and the effective viscosity equation (4.7), he was able to determine the value of N to be 6.6×10^{23}, which is quite a reasonable value for Avogadro's number.

In section 4.4 I will consider this second relation. In particular, I will connect it to the Fluctuation–Dissipation theorem and highlight the essential role played by correlation functions. However, before getting to this, I want to argue that Einstein's viscosity argument provides an explanation for the relative autonomy of the continuum Navier–Stokes equations from lower-scale details of Brownian motion. In other words, I want to argue that Einstein's calculation here can be seen as providing an answer to the question (**AUT**).

4.3.1 Summary: An Answer to (AUT)

As noted, I believe that the theoretical calculation of an accurate value for Avogadro's number (as a result of having derived the effective value for the viscosity of the mixed system, η^*) and the consequent confirmation of the existence of molecules, is the achievement most noted by philosophers. However, the theoretical calculation of that effective value *itself* employs the hydrodynamic method introduced in Chapter 1 and further developed in Chapter 3. We now have a better understanding of how properly to answer the question posed by (**AUT**), which was the focus of attention in Chapter 2. Einstein's derivation of equation (4.7) provides

---

[10]Again, this relation expresses the content of the Fluctuation–Dissipation theorem. I discuss this in more detail in section 4.4. See also (Einstein, 1956, pp. 9–12), (Pais, 1982, pp. 91–92), and (Kubo, 1966, pp. 257–258) for more details.

an answer to how the continuum Navier–Stokes equations are, at least in this instance, relatively autonomous from the lower-scale molecular details of the Brownian motion phenomenon. Let me elaborate.

The continuum equations, as we know, assume that the fluids whose behaviors they govern are just blobs. According to those equations, the fluids have no structure whatsoever all the way down to the infinitesimal. Einstein nevertheless explains how those equations can still yield correct predictions about the behavior of heterogeneous systems like the solute–solvent system definitive of the Brownian motion problem—a system that definitely posits lower (meso-)scale structure—namely, the solute molecules as distinct structures in the solvent.[11]

The explanation for the relative autonomy of the Navier–Stokes equations from the molecular/atomic details proceeds by demonstrating the possibility of finding an effective viscosity for a *fictitious homogeneous* system that satisfies those continuum equations and that *duplicates the behavior of the actual heterogeneous solute–solvent system*. Sometimes this is called an "equivalent homogeneity" (Christensen, 2005, pp. 32–40). Finding such an effective viscosity for an *homogeneous* system is required because (as noted) the continuum equations do not recognize any lower-scale structure whatsoever. In other words, the Navier–Stokes equations only describe the behavior of homogeneous blobs.

Thus, the actual heterogeneous lower-scale details of the solute–solvent are basically irrelevant, as the Navier–Stokes equations are guaranteed to predict the continuum-scale behavior of the actual system once one has the correct material parameter—the effective viscosity, η^*. The lower-scale details are, however, not *absolutely* irrelevant in that they are encoded in that material parameter. The multiscale argument by which η^* was determined is essential to the demonstration of this relative autonomy. It is essential to answering (**AUT**). Einstein's pioneering argument allows us

---

[11] An RVE for this heterogeneous system will reflect this structure at scales above the atomic.

to account for the success of the continuum, phenoménological theory of fluid dynamics.

4.4 Brownian Motion and the F–D Theorem

In section 4.2 I introduced the hydrodynamic equation for Brownian motion, the Langevin equation, by first considering the effect of the random kicks on the particle. Then I considered the friction term. In this section I want to run through the same argument, only this time starting with the friction term. I shall then indicate the role of correlation functions in characterizing the random force term and its connection with the Fluctuation–Dissipation theorem explicitly.[12]

In the linear regime we assume that the friction forces on the particle take the form:[13]

$$m\ddot{x} = -\eta\dot{x} \tag{4.9}$$

This equation has the solution:

$$\dot{x}(t) = \dot{x}(0)e^{-\frac{\eta}{m}t} \tag{4.10}$$

Now, if this were the only force acting on the Brownian particle, the particle's velocity would decay exponentially to zero. In effect, we have the opposite problem from the one we faced in suggesting that the kicks were the only forces at work. Since the particle does not stop moving, we need to add a random force term to equation (4.9) that reflects the kicks from the solvent molecules. So, again we get the Langevin equation (4.2):

$$m\ddot{x} = -\eta\dot{x} + R(t) \tag{4.11}$$

---

[12]The discussion here follows to some extent that found in (Eastman, Chapter 7).

[13]Recall that η is the friction coefficient introduced in equation (4.2).

Let us do a bit more and try to say what the random force term, $R(t)$, actually looks like. As before, we make some assumptions about it:

1. $\langle R \rangle = 0$. The particle has no preferred direction.

2. Stationarity or no preferred time: $\langle R(t)R(t + \delta t) \rangle$ depends only on δt.

3. The kicks are independent of one another for long times: $\langle R(t)R(t + \delta t) \rangle = 0$ if $\delta t > \epsilon$ for some ϵ.

Next we solve the Langevin equation (4.11) to get:

$$\dot{x}(t) = \dot{x}(0)e^{-\frac{\eta}{m}t} + \frac{1}{m}\int_0^t e^{-\frac{\eta}{m}(t-t')}R(t')dt' \qquad (4.12)$$

The first term on the right-hand side is just the friction term we started with. The second represents a sum of kicks, each of which decays in time by a factor weighted by the difference $t - t'$, which is the time since a kick.

We are interested in the long-time behavior of this solution to the Langevin equation. In this limit, the first term in equation (4.12) drops out:

$$\lim_{t\to\infty} \dot{x}(t) = \frac{1}{m}\int_0^\infty e^{-\frac{\eta}{m}(t-t')}R(t')dt' \qquad (4.13)$$

If we now square both sides and take an (equilibrium) ensemble average we get the mean kinetic energy:

$$\lim_{t\to\infty} \langle \dot{x}^2(t) \rangle = \frac{1}{m^2}\int_0^\infty \int_0^\infty e^{-\frac{\eta}{m}(2t-t'-t'')}\langle R(t')R(t'') \rangle dt'dt'' \qquad (4.14)$$

Given assumption 2 above, the correlation function $\langle R(t')R(t'') \rangle$ only depends on the time difference and not the individual time t', so we can rewrite it as $\langle R(0)R(t'' - t) \rangle$.

One further simplification: we make two variable changes, letting $r = (t' - t'')$ and $s = (t'' - t')$, then we have

$$\lim_{t \to \infty} \langle \dot{x}^2(t) \rangle$$

$$= \frac{1}{m^2} \int_{r=0}^{\infty} \int_{s=-\infty}^{\infty} e^{-\frac{\eta}{m}(2r+s)} \langle R(0)R(s) \rangle ds\, dr \qquad (4.15)$$

$$= \frac{1}{2\eta m} \int_{\infty}^{\infty} e^{-\frac{\eta}{m}s} \langle R(0)R(s) \rangle ds \qquad (4.16)$$

Finally, we use the equipartition theorem[14] to relate the average kinetic energy to the absolute temperature:

$$\frac{1}{2} m \langle \dot{x}^2(t) \rangle \rangle = \frac{kT}{2} \qquad (4.17)$$

Cleaning this up yields

$$\langle \dot{x}^2(t) \rangle = \frac{kT}{m} \qquad (4.18)$$

Plugging this into equation (4.16) yields

$$2\eta kT = \int_{\infty}^{\infty} e^{-\frac{\eta}{m}s} \langle R(0)R(s) \rangle ds \qquad (4.19)$$

Assumption 3, above, says that the kicks are uncorrelated or independent of one another except for short times $\delta t < \epsilon$. The timescale for which this assumption is valid is given in equation (4.12). The friction due to the kicks decays by a factor of e in the time $\frac{m}{\eta}$. So if we take $\epsilon \ll \frac{m}{\eta}$ ensuring that the timescale at which the kicks are correlated with one another is much shorter than that at which the friction is observable, we get an expression of the Fluctuation–Dissipation theorem:

---

[14]The equipartition theorem states, roughly, that all energetically accessible degrees of freedom share equally the available energy. Furthermore, it states that each quadratic degree of freedom will, on average, possess an energy $(1/2)kT$, with k being Boltzmann's constant and T being the absolute temperature.

$$2\eta kT = \int_{\infty}^{\infty} \langle R(0)R(s)\rangle ds \qquad (4.20)$$

This equation relates the (auto)correlation function R to the dissipation factor η and the temperature T. In other words, as I have said before, the short-term correlations due to fluctuations are related to the long-term dissipation due to friction. Both are the result of the impacts of the solvent molecules on the solute particles.

4.5   Conclusion

This chapter went into some detail about Einstein's work on Brownian motion from two perspectives that have, I believe, been largely ignored by philosophers of science. Einstein investigated Brownian motion from a mesoscale perspective—one that reaches *up* to the continuum equations of fluid mechanics in calculating the effective viscosity, η^*, of the solute–solvent. In addition, it is from this mesoscale perspective—a perspective that recognizes the solute–solvent as a heterogeneous system—that he derived an expression for what is now called the Fluctuation–Dissipation theorem.

One of Einstein's stated goals was to determine the size (radius) of the Brownian particle. In fact, in the paper entitled "A New Determination of Molecular Dimensions," Einstein lays out his strategy as follows:

> It will be shown now in this paper that the size of the molecules of the solute in an undissociated dilute solution can be found from the viscosity of the solution and of the pure solvent, and from the rate of diffusion of the solute into the solvent, if the volume of a molecule of the solute is large compared with the volume of a molecule of the solvent. For such a solute molecule will behave approximately, with respect to its mobility in the solvent, and in respect to its influence on the

> viscosity of the latter, as a solid body suspended in the solvent, and it will be allowable to apply to the motion of the solvent in the immediate neighbourhood of a molecule the hydrodynamic equations, *in which the liquid is considered homogeneous, and, accordingly, its molecular structure is ignored.* (Einstein, 1956, pp. 36–37, emphasis added.)

One remarkable aspect of this argument is that to determine the molecular size (radius) of a particle (considered, as Einstein did, to be a sphere with no internal structure[15]) *requires the calculation of a viscosity* (a material parameter figuring in the hydrodynamic—Navier–Stokes—equations (4.4)).[16] As he says, in doing so the liquid can be considered homogeneous so that the hydrodynamic equations can apply in the neighborhood of that particle. Note the fact (and it is indeed a fact) that the continuum Navier–Stokes theory is not a fundamental theory, as it does not even recognize molecular or atomic structure. Nevertheless, we have an argument that tells us about *particle size*—structure at a lower scale—that depends on the validity of a *non-fundamental continuum theory* at scales spatially much larger than those of the particle and of the molecules composing the surrounding medium (the solvent).

In providing this argument, Einstein displays how one can answer the question (**AUT**). Recall, this is to explain how systems that are heterogeneous on a micro-scale can nevertheless display the same pattern of behavior on a macro-scale. Answering this provides a justification for our continued reliance on continuum theories to understand the world at macro-scales. That justification involves the demonstration that for the most part, fundamental-scale details are irrelevant.

---

[15]This follows from assumptions 5 and 6 enumerated in section 4.2.

[16]Just to clarify: Einstein's talk of "hydrodynamic equations" here refers to the Navier–Stokes equations (4.4), and not to Kadanoff's and Martin's linearized mesoscale equations.

In deriving a connection between the mean square fluctuations of the Brownian particle and the viscosity, Einstein also provided the first theoretical expression of the Fluctuation–Dissipation theorem. In so doing, he set the stage for hydrodynamic/correlation function methodology that privileges the mesoscale (via the relations between correlation functions and structures in mesoscale RVEs). This methodology was not really appreciated until the late 1950s and early 1960s. See (Kadanoff and Martin, 1963; Kubo, 1966) and references therein.

The next chapter examines how the very same hydrodynamic/correlation function methods, used in the context of a Brownian fluid, are operative in what at first seems to be a completely different context. This the problem of understanding the behaviors of materials such as steel beams and wooden struts.

Chapter 5

From Brownian Motion to Bending Beams

5.1 Introduction

At the beginning of the last chapter I quoted Pais suggesting that Einstein's methods actually have a wide range of applicability. Specifically, he notes that many problems of rheology across different disciplines can be handled following Einstein's lead. In this Chapter I show that not only do his methods bear on fluid problems, but they also provide insight into modeling of material behaviors such as bending beams and electrical and thermal conductivity. Specifically, these methods (and their subsequent refinements) have been employed in materials science. As such, this kind of reasoning really goes *beyond* the hydrodynamic behavior of many-body systems.

5.2 Bulk Properties of Heterogeneous Systems

The problem of determining continuum properties of materials from the "fundamental" microscopic structure is, as noted in Chapter 1, really the age-old problem of reduction in physics. In a 1974 review article entitled "Two-Phase Materials with

A Middle Way: A Non-Fundamental Approach to Many-Body Physics. Robert W. Batterman,
Oxford University Press. © Oxford University Press 2021.
DOI: 10.1093/oso/9780197568613.003.0005

Random Structure," G. K. Batchelor makes a distinction between reductions in physics and those in engineering contexts. It is a distinction that depends on what is taken to be the (fundamental) microstructure:

> If by microscopic structure we mean the arrangement and motion of molecules, the problem may be called, according to the context, kinetic theory, liquid state theory, metal physics, or crystallography. But in some circumstances, there exists a structure scale large compared with the molecular dimensions and at the same time small compared with the overall dimensions of the given sample of material, and the term *microscopic structure* may then refer to the arrangement and properties on this intermediate scale. In these circumstances, in which the material is still a continuum when viewed on the microscopic scale, the problem moves from the domain of the physicist to that of the engineer, and is associated with the terms heterogeneous media, multiphase flow systems, and composite materials. (Batchelor, 1974, p. 277)

In the physics "domain," as Batchelor sets it up, one aims *directly* to connect molecular and atomic structures to continuum or macroscopic behaviors. Sometimes such direct connections are realizable. Usually these cases are homogeneous. For example, one can upscale from kinetic theory to determine the ideal gas law. But, as Kadanoff and Martin have argued, such direct connections are almost always impossible to achieve. Claims that direct connections are always possible evoke Truesdell's "real" physicists who insist that realizing the direct connection is *merely* a matter of mathematics. Likewise, Sober claimed, on behalf of reductionists in the multiple realizability debate, that such connections were "in principle" possible. In what follows I will be arguing that the engineering domain (or an engineering approach to understanding theories of many-body systems) is much more fruitful and important than the "fundamental" bottom-up approach of the "in principle," "real" physics crowd.

Batchelor's engineering approach, despite surface differences, is actually an instance of the approach to complex many-body systems proposed by Kadanoff and Martin. In fact, as I noted, Einstein's work on Brownian motion can be seen as fitting in with both of these programs. We saw how the Brownian motion fits within the hydrodynamic approach in the last chapter; specifically, in the context of employing the Fluctuation–Dissipation theorem. In this chapter we examine the general problem of determining bulk (continuum) properties from lower-scale structure largely from a materials science (or an engineering) perspective.

An important intersection between the hydrodynamic methodology and the materials science methods for determining or predicting bulk properties of heterogeneous materials is the focus on "transport" properties of conserved quantities. The aim is to determine the ability of a heterogeneous material "as a whole to transfer some conservable quantity, such as heat, mass, electric charge, or momentum, in response to an imposed gradient of intensity of that conservable quantity" (Batchelor, 1974, p. 277). Furthermore, these methods also assume that small departures from equilibrium will give rise to a flux of the conserved (conservable) quantity that will be approximately *linear* in that near- but out-of-equilibrium situation.

Following Batchelor, we can introduce some notation to talk about transport properties. Consider first an homogeneous system, e.g. a thermal conductor (a copper plate maybe), in thermal equilibrium. We might be interested in how a small disturbance by a temperature gradient $\mathbf{G}$ affects (or induces) a heat flux $\mathbf{F}$ in the plate. In general, in this weakly perturbed equilibrium state we have the following equation:

$$\mathbf{F} = \mathbf{k} \cdot \mathbf{G} \qquad (5.1)$$

where $\mathbf{k}$, a constant of proportionality, is the transport coefficient characteristic of the medium (in this case the heat conductivity of copper).[1] Recall again the previous reference to Hooke's law for a spring. The transport coefficient, which is different for different materials, is the analog of the spring constant in this more general context.

Of course, the more interesting cases concern materials that are heterogeneous at some lower "microscopic" scale but one which is still significantly above the atomic lattice scale. As noted, we treat the heterogeneities at this intermediate scale as continua, because from the perspective of the lattice scale, that is how they appear. In many instances we are interested in materials that are heterogeneous in this sense, but that are also statistically homogeneous. In other words, these are materials for which it is possible to define a representative volume element or RVE that statistically represents the heterogeneous material distribution in any selected, actual volume element.[2] For the purposes of this discussion we will assume we are dealing with material mixtures of two disperse components. That is, in the RVE there will be a matrix material that contains inclusions of a second material or phase. The analog of equation (5.1) for this heterogeneous case is

$$\langle \mathbf{F} \rangle = \mathbf{k}^* \cdot \langle \mathbf{G} \rangle \tag{5.2}$$

where $\mathbf{k}^*$ is now an *effective* transport coefficient characterizing the macroscopic behavior of the material, exactly like Einstein's η^* characterized the viscous or frictional macroscopic/continuum behavior of the solute–solvent mixture. The brackets, as usual, indicate averages—in this case volume averages. The theoretical goal is to determine the value for the effective transport coefficient; just as one of Einstein's goals was to determine the effective viscosity of the solute–solvent mixture—also a disperse two component mixture.

---

[1]If the conserved quantity is a scalar quantity (as is heat), then $\mathbf{k}$ is a second-order tensor. (If $\mathbf{G}$ is the gradient of a conserved *vector* quantity like momentum, then $\mathbf{k}$ will be a fourth-order tensor.)

[2]Recall the composite RVE from Chapter 1, figure 1.4.

Type of System	Quantity $\mathbf{F}$	Quantity $\mathbf{G}$	Coefficient $\mathbf{k}^*$	Differential Equation For Each Phase
Thermal Conductor	Heat Flux	Temperature Gradient	Thermal Conductivity	$\mathbf{F} = \mathbf{k} \cdot \mathbf{G}$ $\nabla \cdot \mathbf{F} = 0$
Electrical Conductor	Electric Current	Electric Field	Electrical Conductivity	$\mathbf{F} = \mathbf{k} \cdot \mathbf{G}$ $\nabla \cdot \mathbf{F} = 0$
...	...	...	...	
Small Fluid or rigid particles suspended in incompressible Newtonian fluid	Deviatoric Stress	Rate of Strain	Shear Viscosity	$\mathbf{f} = 2\mu\mathbf{G}$ $\nabla \cdot \mathbf{F} = 0$
Elastic inclusions embedded in an elastic matrix	Stress	Strain	Rigidity and Bulk moduli	$\mathbf{F} = 2\mu\mathbf{G} + \lambda\mathbf{G}\mathbf{I}$ $\nabla \cdot \mathbf{F} = 0$ μ, λ are local Lamé constants

Table 5.1: Different Two Phase Upscaling Problems (Batchelor, 1974, p. 229)

Batchelor provides a table of different cases of "two phase disperse systems with random structure in which the problem is to determine an effective transport coefficient" (Batchelor, 1974, p. 229). These are systems that are mixtures of two materials or phases, each one of which is taken to be homogeneous. One phase will be discrete inclusions or particles, and the other will be a connected matrix whether fluid or solid. Thus, the heterogeneity is present in the mesoscale—any structure within each phase is ignored. Table 5.1 lists some of the two phase transport problems that Batchelor groups together within a common theoretical framework. Some of the problems, he notes, display mathematical differences in the governing equations and in boundary conditions that obtain at the interface between the matrix material and the inclusions. Nevertheless:

> [o]nce the principles have been grasped, the common features of the various transport problems become apparent. For the most part, research on the different types of transport problem has proceeded independently, and with the benefit of hindsight we can now see how unnecessary this was. (Batchelor, 1974, p. 229)

Let us consider the problem in the last row. This is the problem we face if we are interested in understanding the bending of beams.[3] We focus on the nature of the stresses that are experienced by the beam upon its being strained. (Stress is the deforming force applied to the material per unit area; whereas, strain is the change in shape, or displacement of the material as a result of the stress.) In this case $\langle \mathbf{F} \rangle$ is the average stress and $\langle \mathbf{G} \rangle$ is the average strain. Normal notation for these quantities are $\langle \sigma_{ij} \rangle$ for average stress and $\langle \epsilon_{ij} \rangle$ for average strain.

---

[3]On the assumption, of course, that the beam is composed of a two-phase disperse material.

Most generally, for a given RVE of volume V, these are defined as follows:

$$\langle \sigma_{ij} \rangle = \int_V \sigma_{ij}(x_i)dv \qquad (5.3)$$

$$\langle \epsilon_{ij} \rangle = \int_V \epsilon_{ij}(x_i)dv \qquad (5.4)$$

The effective linear stiffness or rigidity (the $\mathbf{k}^*$) is given by the tensor $\mathbf{C}^*_{ijkl}$ which is defined by the relation

$$\langle \sigma_{ij} \rangle = \mathbf{C}^*_{ijkl}\langle \epsilon_{ij} \rangle \qquad (5.5)$$

So to determine the value for the effective rigidity, $\mathbf{C}^*_{ijkl}$, we "merely" have to solve equation (5.5) by evaluating equations (5.3) and (5.4).[4] However:

> [a]lthough this process sounds simple in outline, it is complicated in detail, and great care must be exercised in performing the operation indicated. To perform this operation rigorously, we need exact solutions for the stress and strain fields σ_{ij} and ϵ_{ij} in the heterogeneous media. (Christensen, 2005, p. 35)

In fact, determining the exact solutions for the stress and strain fields (appearing under the integrals in (5.3) and (5.4)) requires an infinite amount of information. Consider an RVE that looks something like that in figure 5.1. Of course, one very important bit of information is the volume fraction ϕ of the inclusions. (Recall the importance of ϕ for Einstein's relation determining the effective viscosity in the Brownian motion case.) But it is quite clear that if that is all the information we have in trying to specify the fields σ_{ij} and ϵ_{ij} before performing the volume averaging in equations (5.3)

---

[4]The claim that all that stands between us and a solution to equation (5.5) is "mere" mathematics is the same as those asserted by Truesdell's "real physicists" and by Sober in (Sober, 1999).

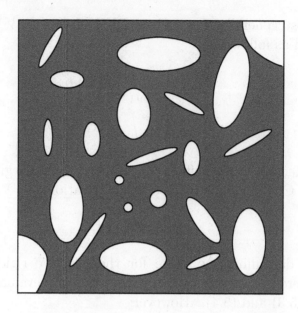

Figure 5.1: Matrix and inclusion (RVE).

and (5.4), we will probably grossly under- or over-estimate the rigidity of the beam. Suppose that the matrix material (dark) in figure 5.1 is much stiffer than the inclusions. And suppose as well that the volume fraction of the inclusions is the same as that of the matrix; i.e., this is a 50–50 mixture. Then upon performing the simple volume averaging we will conclude that the material is much less stiff than it actually is. This is because that averaging just takes the volume fraction into account and ignores the fact that topologically the inclusions are disconnected and the matrix is connected. If one takes these facts into consideration, it is clear that the stiffness of the matrix material will dominate that of the inclusions, and so our 50–50 prediction of the value for $\mathbf{C}^*_{ijkl}$ will be extremely underestimated.

To remedy this, we need more information about the geometry and topology of the structures in the RVE. It is here, as noted in Chapter 1, that correlation functions come into play. In Chapter 3 we focused on time–time correlation functions that allow us to characterize structures like density gradients that exist at some times but then may decay as the

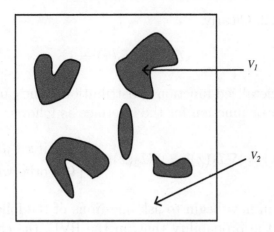

Figure 5.2: Two-phase random medium. (After (Torquato, 2002).)

system relaxes back to an equilibrium state. In the context of materials, the situation appears much more static. We subject our RVE to a strain and ask about the nature of the induced stresses, but we do not (at least not initially) ask if the shapes and locations of the inclusions in the RVE change.

Therefore, in the materials context we need to introduce correlation functions that will help us determine the geometric and topological features present in the RVEs. In this context those correlation functions are best thought of as probability density functions.

Consider figure 5.2 where the inclusions are the regions V_1 and the matrix is the region V_2. Let us further suppose that this two-phase medium is statistically isotropic (essentially that the material looks the same statistically under rigid body rotations). Clearly, if V is the RVE, we have $V = V_1 \cup V_2$ and $V_1 \cap V_2 = \emptyset$. Furthermore, let ∂V be the boundary or interface between V_1 and V_2. If x is a point in the RVE, we can define characteristic (or indicator) functions

$$\mathcal{I}_i(x) = \begin{cases} 1, & \text{if } x \in V_i \\ 0, & \text{otherwise} \end{cases} \tag{5.6}$$

for $i = 1, 2$. Clearly,

$$\mathcal{I}_1(x) + \mathcal{I}_2(x) = 1 \tag{5.7}$$

Using generalized functions (distributions) one can also define an indicator function for the interface as follows:

$$\mathcal{M}(x) = |\nabla \mathcal{I}_1(x)| = |\nabla \mathcal{I}_2(x)| = \begin{cases} 1, & \text{if } x \in \partial V \\ 0, & \text{otherwise} \end{cases} \tag{5.8}$$

We can now begin to ask questions of the following form. What is the probability that, in the RVE, the characteristic function $\mathcal{I}_1 = 1$? That is,

$$Pr(\mathcal{I}_1(x) = 1) = ? \tag{5.9}$$

Let us normalize the probability functions to the RVE so that

$$Pr(x \in V) = 1$$

Then the answer to the question posed by equation (5.9) is just the volume fraction of the inclusions, V_1. Roughly, one can think of throwing a dart randomly at the RVE (where you are a bad shot—random selection—but where you are good enough to be guaranteed to hit a point somewhere in V). We then record the frequency that the dart lands in the inclusion phase. (Say 1 if it is in the the phase, and 0 if it is not.) Thus, this probability function can be thought of as a one-point correlation function, S_1. The more darts one throws, the closer the S_1 will be to representing the volume fraction of V_1. One can get more information by throwing line segments of length r at the RVE for various values of r. Thus a two-point correlation function, $S_2(r)$, will have a value of 1, if both endpoints lie in the inclusion phase, and a value of 0 if one or both lie in the matrix. If one does this for all values of r, then we learn the degree to which two points are correlated with one another in the sense that they both lie in the same phase. One can generalize to throwing triangles of sides (r, s, t) for all values of r, s, and t to gain even more information about the nature of the inclusion phase. And so on for n-point correlation functions. The idea is that by throwing our darts, line segments, triangles, ..., and taking the limit relative frequency of 1s for each correlation function S_i, we can reconstruct the geometric and topological properties of the material RVE.

Torquato (2002), in chapter 2 of *Random Heterogeneous Materials: Microstructure and Macroscopic Properties*, makes this all precise. In fact, much more can be done. One can ask, for example, about the probability that a line segment of length r lies entirely in a given phase, and one can even define surface correlation functions to learn about the nature of the interface between the different phases, ∂V.[5] Full information about the structures in the RVE and the exact values for σ_{ij} and ϵ_{ij} would require n-point correlation functions with $n \to \infty$.

The general lesson from this discussion is that very similar methods apply to what *prima facie* appear to be quite distinct problems—determining the effective viscosity of a solute–solvent (in the Brownian case), and determining effective stiffness coefficients for elastic materials. All involve, in an essential way, a mesoscale approach in which correlation functions play a critical role in characterizing (to the extent that they can) the relevant geometric and topological (shape) details of the RVEs beyond the simple volume fraction.

It remains to say a bit more about the differences between the "flowing" problems like Brownian particles and the spin-diffusion problem discussed in Chapter 3, and the "static" problems of thermal conductivity and elastic material behavior. For two-phase disperse system problems that are stationary, like thermal conductivity and bending beams, the geometry of the interface is most often determined by the

---

[5]These latter are not probability functions, but they can be determined by assigning the interface a finite thickness and then examining the limiting behavior as the thickness goes to zero (Torquato, 2002, pp. 43–44).

way the material was manufactured (whether in a factory or by nature). Time-independent, n-point correlation functions/probability distributions can provide us with statistical information about the structures in the RVE, as we have just seen.

> But in the [cases where the interface geometry is changing/flowing], in which the particles are in continual motion, the statistical properties of the interface may be determined in part by the initial conditions of manufacture and in part— often wholly, in practice—by the imposed bulk motion and the associated flow processes; in other words, determination of the statistical geometry of the interface is here a part of the problem, and we may expect "history" and nonisotropic-structure effects to turn up in bulk properties of two-phase systems whose local transport properties give no hint of the existence of such effects. (Batchelor, 1974, p. 230)

Batchelor's invocation of "history" here may naturally be taken to refer to correlations/fluctuations and their decay over time. These correlations and fluctuations are, of course, just those that play the central role in the hydrodynamic approach to many-body systems promoted by Kadanoff and Martin.

5.3 Conclusion

Batchelor distinguishes two kinds of reduction programs or upscaling problems according to different posits about the nature of the fundamental microstructure in a given situation or system. On the one hand, if microstructure consists of atoms or molecules, we are in the "physics" domain. On the other hand, if microstructure refers to mesoscale structures in between the scale of atoms and molecules and continuum scales, then we are in the "engineering" domain. The former upscaling/reduction problem aims to make *direct* connections between the atoms or molecules and the continuum properties and behaviors of many-body systems.[6] The latter upscales from continuum mesostructures to continuum properties and behaviors. Setting aside the physics domain problems, this chapter has aimed to demonstrate deep affinities between hydrodynamic descriptions of "flowing" problems, specifically Brownian motion and the spin-diffusion example from Chapter 3, and *static* problems that appear in more conventional engineering and materials science contexts.

I think that a lesson to be learned from this discussion, and from the discussions in earlier chapters, is that the dichotomy between a domain of "physics upscaling problems" and a domain of "engineering upscaling problems" is not very well motivated. It may be natural from a disciplinary perspective (one that distinguishes engineering departments from physics departments), but from a perspective that focuses on scientific methodologies there is not much to be said in favor of the dichotomy. Direct connections between fundamental microstructure and continuum properties that define the physics domain are extremely rare. Paradigm examples come from some successes of kinetic theory that are possible only if the microstructure is homogeneous so that limiting volume averaging is a viable mathematical strategy. *The example of the 50–50, two-phase mixture* (see figure 5.1) *demonstrates the failure of limiting volume averaging strategies for any heterogeneous mixture.* Once one recognizes that most upscaling problems involve complex systems with heterogeneous structures at lower scales, it becomes clear that such a simple strategy is doomed to fail.

---

[6]This is the "natural" approach that follows from thinking that the aim of physics is to start from *the* fundamental description of a system's components and from that derive upper-scale, phenomenological behaviors and properties. As noted in Chapter 1, this approach may be natural from the point of view of Hilbert's 6th problem, but it is unnatural from the field-theoretic point of view advocated by Schwinger, Kadanoff, and Martin.

While there are some differences between the (fluid) Brownian motion problem tackled by Einstein and the (static) materials science problems such as determining the stiffness coefficients for a composite beam or strut, the methodologies are, in fact, really the same. All involve a focus on mesoscale RVEs, hydrodynamic descriptions, and correlation functions.

The next chapter considers the prospect of elevating an "engineering" mesoscale-centric approach to many-body systems as the preferred methodology over a bottom-up "physics" approach.

Chapter 6

An Engineering Approach

6.1 Introduction

In the previous chapter we encountered Batchelor's distinction between upscaling problems in the domain of "physics" and those in the domain of "engineering." If we adopt, as is suggested by Batchelor, a narrow reading of "physics domain problems" to refer to those that directly try to upscale from the lowest fundamental structure to continuum structures, then the two domains are quite distinct. However, if we take a broader view of the domain of physics to include the hydrodynamic descriptions of condensed-matter theorists and the homogenization strategies of materials scientists, then the domains simply reflect the fact that physics departments and engineering departments are often found in different campus buildings.

That said, I think by reflecting a bit further on Truesdell's caricature of "real" physicists and comparing their methods to those of actual working condensed-matter physicists and to those of material scientists, we may gain some further insight into the mesoscale methods that we have focused on so far. One motivation for this comparison of methods comes from a seemingly unlikely source: Nobel Laureate Julian Schwinger.

A Middle Way: A Non-Fundamental Approach to Many-Body Physics. Robert W. Batterman, Oxford University Press. © Oxford University Press 2021.
DOI: 10.1093/oso/9780197568613.003.0006

In the 1960s, Schwinger (and the particle physics community as a whole) faced severe problems when trying to construct a theory of strong interactions. In his Stanley H. Klosk lecture at NYU (published as "Julian Schwinger's Engineering Approach to Particle Theory"), Schwinger noted that

> [h]istorically, relativistic quantum mechanics had proved very successful in explaining atomic and nuclear physics until we got accelerators sufficiently high in energy to create the strongly interacting particles, which include particles that are highly unstable and decay through very strong forces. The ordinary methods that had evolved up to this point were simply powerless in the face of this new situation. (Schwinger, 1969, p. 19)

It seemed there was no fundamental theory to which theorists could appeal using the perturbative techniques that worked so well in quantum electrodynamics. There were, *prima facie*, two options. One might continue to try to develop an operator field theory (à la QED) that could handle the strong interactions in spite of the fact that the perturbative techniques were of no avail. This would involve, he said, replacing "the particle with three-dimensional space itself" (Schwinger, 1969, p. 19). Such a view demands "that one is indeed able to describe physical phenomena down to arbitrarily small distances, and, of course, that goes far beyond anything we know at the moment" (Schwinger, 1969, p. 19). In fact, he noted that this approach involves a host of speculations about "how particles are constructed before you can even begin to discuss how particles interact with each other"[1] (Schwinger, 1969, p. 19).

---

[1]In fact, by the 1960s Schwinger was skeptical of the operator field approach even in the context of QED (the theory he invented!), despite the major successes of that theory. QED itself demands that one be "able to describe physical phenomena down to arbitrarily small distances." In retrospect, QED worked because the weakly coupled fundamental fields

In stark contrast, one could adopt a much more empiricist/phenomenological approach—S-matrix theory—and insist that there is

> nothing more fundamental than particles and that, when you have a number of particles colliding with each other and the number of particles ceases to be constant, all you can do is correlate what comes into a collision with what goes out, and cease to describe in detail what is happening during the course of the collision. (Schwinger, 1969, p. 22)

Schwinger's suggestion was to opt for a *third way*—one that eschewed both the fundamental operator field theory approach and the phenomenological S-matrix approach. Rejecting the former would be to give up on the approach that a "real" physicist would take: he would find the fundamental theory and then "simply" solve for results that connect up with experiment. Schwinger, however, asserted that "[t]he true role of fundamental theory is not to confront the raw data, but to explain the relatively few parameters of the phenomenological theory in terms of which the great mass of raw data has been organized" (Schwinger, 1969, p. 19). He called this an "engineering approach to particle theory" and argued that "[w]hat we should *not* do is try to begin with some fundamental theory and calculate" (Schwinger, 1969, p. 19).

Note here that Schwinger's focus is on the few parameters that we employ to organize the world. These are the material parameters and order parameters that, as I have noted, code for lower-scale detail in various ways. They are the objects of the hydrodynamic and materials science methods upon which we have been focusing. In section 6.4 I will also make a brief

---

of the theory just happened to be those associated with the particles one observes in the detectors. In the case of the strong interactions (QCD), the measured particles' properties do not reflect the properties of the fundamental fields, because of color confinement. Thanks to Porter Williams for helping me get straight about some of Schwinger's ideas here.

case for the fruitfulness of an engineering, mesoscale approach to some biological modeling.

Of course, Schwinger's main interest was particle theory and, specifically, attempts to understand the strong interactions. He worked for many years on what he called "source theory"—the theory that he hoped would realize this engineering approach to particle physics. It is fair to say that Schwinger's source theory did not catch on in the particle physics community.

6.2 Schwinger's Engineering Approach

The motivation behind Schwinger's engineering approach to particle theory arose, in part, from work he did on waveguides during the Second World War. He recognized, rather quickly, that it was a mistake to try to design waveguides using a bottom-up approach that started with the fundamental equations of electromagnetism—Maxwell's equations. The main reason for eschewing this bottom-up "physics domain" approach was because

> [t]he fundamental theory is too complicated, generally too remote from the phenomena that you want to describe. Instead, there is always an intermediate theory, a phenomenological theory, which is designed to deal directly with the phenomena, and therefore makes use of the language of observation. On the other hand, it is a genuine theory, and employs abstract concepts that can make contact with the fundamental theory.
>
> (Schwinger, 1969, p. 19)

He notes that focusing on the intermediate theory leads to the kind of phenomenological theory that engineers often use. Such a theory "can be connected to the fundamental theory at one end, and at the other it is applied directly to the experimental data" (Schwinger, 1969, p. 19). In the context of designing waveguides, Schwinger noted that fundamental theory—Maxwell's theory in this case—only plays a role in theoretically determining the values of a "few parameters, the effective lumped constants that characterize the equivalent circuits" (Schwinger, 1969, p. 19).

Here again we see a reference to effective parameters or "lumped constants" that reflect an important intermediate or mesoscale. These are the abstract concepts that "can make contact with the fundamental theory." They are the analogs, in the context of waveguide design, of the material parameters and order parameters we have encountered in our earlier discussions. Their "contact with the fundamental theory" is of the same nature as the connections with lower scales in the examples from the previous chapter. That is to say, the "lumped constants" code for underlying details in the sense that they represent structures (inductors, capacitors, i.e., circuit components) in a mesoscale RVE that allow the determination of the continuum, Maxwell solutions for electromagnetic behavior. (See (Marcuvitz and Schwinger, 1951) for details.)

The alternative (engineering) approach to particle theory was also motivated by a desire to avoid rampant speculation. On the one hand, the field-theoretic approach demanded rethinking the very nature of particles, reducing them to regions of spacetime. On the other hand, the *S*-matrix approach requires speculating that particles are both unanalyzable and "are made from nothing but themselves" (Schwinger, 1969, p. 22). Instead, he says

> [w]e want to eliminate speculation and take a
> pragmatic approach. We are not going to say that
> particles are made out of fields, or that particles
> sustain each other. We are simply going to say
> that particles are what the experimentalists say
> they are. But we will construct a theory and not
> an experimentor's [sic] manual in that we will look

at realistic experimental procedures and pick out
their essence through idealizations.

(Schwinger, 1969, p. 22)

This way of describing the engineering approach appears
manifestly pragmatic.[2] Schwinger himself even characterizes
it that way in the previous quote. However, I want to argue
that despite this apparent pragmatic motivation, mesoscale
approaches are, in fact, often motivated by much more than
a desire to avoid speculation. We can see this most clearly
if we focus on the goals of upscaling and of explaining the
relative autonomy of our continuum theories from lower-scale

---

[2]However, in a lecture entitled "Fields and Particles," Schwinger
(1963) himself offers a temporal scale critique of the very possibility of
reducing particles to spacetime regions. He argues that the very concept
of a particle will not actually make sense at the scale of spacetime points
or regions. So, at least here, he seems to think there are non-pragmatic
reasons to eschew starting with fundamental theory.

> In a sense, the concept of a physical particle is one which
> involves sufficient time for its properties to be measurable
> with sufficient accuracy. For short time descriptions
> obviously the notion of the physical particle becomes
> inapplicable. If one asserts that a more detailed description
> should be possible in seeking for something simpler, more
> fundamental than the particles as we observe them, then
> it is not in terms of the physical particles that you must
> look for some simpler, more fundamental description. Now
> the decomposition of the total energy and momentum
> into the additive contributions of the particles is erased.
> Again we look for an analysis in terms of something
> simpler, something additive. But what is now introduced
> is the idea of the decomposition of energy and momenta
> into the contributions of small volume elements of three-
> dimensional space; the idea that the various volume
> elements of space support energy and momentum, that
> the energy-momentum contributions are additive, that the
> analysis in question is a space-time analysis, and that the
> concept of space-time representations or descriptions is
> meaningful down to arbitrarily small distances. Whether,
> in fact, this extreme point of view is an inherent part of the
> field ideas, I shall not insist on. (Schwinger, 1963, pp. 10–11)

"fundamental" details. We have seen the indispensability of mesoscale structures in these contexts in Chapters 4 and 5. Mesoscale correlational structures are necessary for (theoretically) determining values for continuum parameters that appear in our upper-scale theories. And, as I have argued, being able to perform these upscaling tasks allows for the explanation of the relative autonomy of the continuum theories—answering (**AUT**).

There are analogies here to recent attempts to interpret effective quantum field theories in realist terms (Williams, 2019). While not directly aiming to connect empirically adequate low-energy effective physics to a more "fundamental" high-energy theory,[3] one goal, as Williams says, is to demonstrate that "the fundamental structure of a quantum field theory can be in many respects irrelevant for answering why-questions about its behaviour at longer distances" (Williams, 2019, p. 228).

Williams further denies that treating the ontological commitments of an effective QFT realistically restricts those commitments to the theory's observable content (Williams, 2019, p. 232). I agree. In the present context, the analog of that "observable" content are the everyday continuum-scale observations of how, for example, beams bend under loads. As emphasized by Schwinger, the intermediate theory is a *genuine* theory and not an experimenter's manual. The theoretical commitments of our engineering approach to upscaling are the correlational structures that are indiscernible at the "fundamental" scale but are required to determine the effective parameters that appear in the continuum laws and descriptions.

There is an *apparent* tension between my claims that we should treat the mesoscale structures realistically, and my talk of finding a "fictitious homogeneous system" that displays the same continuum behavior of the actual heterogeneous system. The discussion of Einstein's upscaling to determine

---

[3]The reason for this is simple. Unlike in the cases considered in this book, we do not know what the more "fundamental" theory is in the context of QFT.

the effective viscosity of the solute–solvent in Brownian motion was one example. In section 4.3.1 I also noted that this fictitious system is sometimes called an "equivalent homogeneity" (Christensen, 2005). *Prima facie*, this talk of fictions seems manifestly instrumentalist and largely pragmatic.[4]

However, I do not think it is. Recall that the continuum equations—Navier–Stokes for fluids, Navier–Cauchy for solids—manifestly get the ontologies of their subjects wrong. Despite this, they are remarkably successful, safe, robust, and empirically adequate at continuum scales. In pointing out that the goal of upscaling from mesoscales is in part to explain these features, I am not treating the equivalent homogeneities or fictitious systems instrumentally. Instead, I am demonstrating how a model that includes *none* of the actual heterogeneous structure can nevertheless possess these remarkable properties. Neither is this to treat the equivalent homogeneity (or fictitious homogeneous system) realistically. *It is, however, to treat the mesoscale correlational structures realistically.* And, as noted, this allows us to explain the remarkable and continued success of the continuum theories. Put another way, this shows how the ontologically correct mesoscale model can reproduce the "fictitious" long-distance (continuum-scale) model. Furthermore, in demonstrating this I show that the "fundamental" microstructure really contributes only minimally to the long-distance behaviors, thereby both answering (**AUT**) and justifying the continued use of the upper-scale continuum theory.

In the next section I will argue further that the engineering approach is an important approach to many-body systems. The parameters—order parameters and material parameters—genuinely allow for the representation of such systems in ways that bottom-up approaches do not.

6.3 Order Parameters, Mesoscales, Correlations

In this section I argue that mesoscale properties are necessary for understanding the inter-theory relations between

---

[4]Thanks to Colin Allen for pushing me on this point.

"fundamental" theories and non-fundamental (continuum) theories. Such mesoscale features allow for the connection Schwinger emphasizes between experimental data (from continuum-scale experiments) and the lower-scale "fundamental" theory.

To begin, let us recall the discussion in Chapter 3 where we considered the spin diffusion equation for a large collection of spin-carrying particles. I constructed the magnetization, $M(\mathbf{r}, t)$, which is a function of the density of up- and down-spins at a spacetime point $(\mathbf{r}, t)$. This was equation (3.4), which I repeat here:

$$M(\mathbf{r}, t) = \gamma \left(\rho_+(\mathbf{r}, t) - \rho_-(\mathbf{r}, t) \right)$$

where $\rho_\pm$ are the densities in the up/down directions at the point $(\mathbf{r}, t)$ and γ is the spin magnetic moment of each particle. This was the conserved quantity of interest in that example. $M(\mathbf{r}, t)$ is an order parameter. If one plots its values as a function of temperature, we can easily see in what sense it describes an ordered state of the system. One can see in figure 6.1 that the magnetization $M(\mathbf{r}, t)$ is equal to 0 when

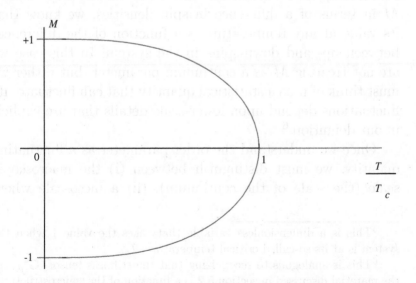

Figure 6.1: Magnetic order parameter. (After (Selinger, 2016).)

the reduced temperature,[5] $\frac{T}{T_c}$, is greater than 1. For those values, the thermal energy has randomized the spins, and so the difference between $\rho_+(\mathbf{r}, t)$ and $\rho_-(\mathbf{r}, t)$ is 0. At criticality, $\frac{T}{T_c} = 1$, and the value of $M(\mathbf{r}, t)$ suddenly becomes non-zero either in the "up" or "down" direction. Above $\frac{T}{T_c} = 1$, the system has complete rotational symmetry—it looks the same in all directions. But below $\frac{T}{T_c} = 1$, there is a preferred direction—the direction determined by the greatest number of aligned spins. In the jargon, this is to say that the rotational symmetry is spontaneously broken as the system is cooled past its critical temperature.

If we had come upon our spin system (a lump of ferro-magnetic iron) while hiking with a compass (remember those things?), and the system was below its critical temperature, our compass would tell us that the system had non-zero net magnetization and in what direction. Were we then to proceed to heat up the iron (over our campfire), we could (theoreti-cally) notice that above some temperature the magnetization went to zero. At this level or scale of observation, we would simply think of the quantity M as a *continuum* parameter that describes an interesting aspect of the system and its relationship to temperature. But since we actually defined M in terms of a difference in spin densities, we know that its value at any temperature is a function of the differences between up- and down-spins in the system. In this case we are not treating M as a continuum parameter, but rather we must think of it as a statistical quantity that can fluctuate. Its fluctuations depend upon lower-scale details that are explicit in our definition.[6]

Once we understand the order parameter as a fluctuating quantity, we must distinguish between (i) the macroscopic scale (the scale of the continuum), (ii) a mesoscale where

---

[5]This is a dimensionless variable that takes the value 1 when the system is at its so-called critical temperature, T_c.

[6]This is analogous to recognizing that the stiffness tensor, $\mathbf{C}^*_{ijkl}$, of the material discussed in section 5.2 is a function of the "microstructure" in the relevant representative RVE.

fluctuations in aggregates of atomic scale properties are important, and (iii) the atomic scale where what matters is the detailed natures of the atoms/molecules and spins. The order parameter lives in the intermediate regime.

Michael Fisher (1998) has stressed this point in a review article entitled "Renormalization Group Theory: Its Basis and Formulation in Statistical Physics." He notes that the idea of an effective field-theoretic approach to many-body systems probably has its roots in the work of Lev Landau. Fisher argues that Landau introduced (in fact he says Landau *invented*) the concept of an order parameter (Fisher, 1998, p. 654). As with the material parameters we have mostly been considering, order parameters allow one to say a considerable amount about the macroscopic nature of many-body systems without having to have detailed knowledge about the nature of the system at the most fundamental scale. Fisher believes that the latter way of understanding order parameters yields a profound conclusion about the structure of the world:

> Significantly, in my view, Landau's introduction of the order parameter exposed a novel and unexpected *foliation* or level in our understanding of the physical world. Traditionally, one characterizes statistical mechanics as directly linking the *microscopic* world of nuclei and atoms (on length scales of 10^{-13} to 10^{-8} cm) to the *macroscopic* world of say, millimeters to meters. But the order parameter, as a dynamic, fluctuating object in many cases intervenes on an intermediate or *mesoscopic* level characterized by scales of tens or hundreds of angstroms up to microns (say, $10^{-6.5}$ to $10^{-3.5}$ cm). (Fisher, 1998, p. 654)

The definition of the order parameter, $M(\mathbf{r}, t)$, demonstrates that the parameter is actually coding for correlations between the spin components. Fluctuations in aggregates of the atomic components (the spins) lead to correlations. Thus the value of $M(\mathbf{r},t)$ will be non-zero if spins are more likely to be pointing in the same direction than not. Above the critical temperature, the spins are not correlated; below the critical temperature, they are.

Now, let us reconsider the Fluctuation–Dissipation theorem. It asserts a deep connection between the response of a system to an external push or disturbance and the internal fluctuations of the system in the absence of the push. These internal fluctuations are encoded in correlation functions. Order parameters, as just noted, reflect those correlations. Therefore, in a sense, *pace* Fisher, it is *not* so surprising that order parameters reflect the importance of a mesoscale intermediate between the continuum scale and the atomic scale. The Fluctuation–Dissipation theorem guarantees there will be correlations and density fluctuations at the mesoscale. *The Fluctuation–Dissipation theorem for many-body systems is really a profound result with ontological consequences.* It tells us that we need to take mesoscale structures and the parameters (material parameters or order parameters) that code for those structures via correlation functions to be genuine features of the world. It is in this sense that order parameters intervene or mediate at mesoscales.

"Intervention" and "mediation" are loaded (and apparently causal) terms.[7] In the previous quote, Fisher uses the term "intervene." However, my use of these terms and my understanding of Fisher's use of "intervene" is not intended to have any causal connotations. Instead, I mean these terms to reflect the importance of mesoscales for realizing connections between upper-scale theories and their lower-scale, more "fundamental," counterparts. The importance of the "interventions" or "mediations" is that they allow for a proper understanding of these inter-theory relations. Direct, reductionist, upscaling attempts completely miss the role of the mesoscale in these inter-theoretic connections.

Thus, the engineering approach to many-body systems emphatically is *not* motivated only by pragmatic considera-

---

[7]Again, thanks to Colin Allen for stressing this.

tions. The mesoscale properties that connect the "fundamental" theory to the experimental data (often available only at considerably higher scales) are genuine features that are necessary for properly relating "fundamental" theories to non-fundamental theories. A focus on the mesoscale features also allows for an understanding of the relative autonomy of the non-fundamental theory from the fundamental. The appropriate upscaling relation is, to use Batchelor's classification of fundamental microstructure, a problem in the engineering domain—microstructures at mesoscales.

Furthermore, this suggests that a "middle-out" or "middle-first" strategy for modeling many-body systems is appropriate, and even necessary, for investigating many-body systems. There are theoretical reasons that privilege such a strategy. In Chapter 7 I elaborate on this further. But before getting to this, it is interesting to note that biologists also have recognized the importance of middle-out approaches. In the next section I consider one biological example—a middle-out strategy for modeling bone in humans.

6.4 Multiscale Modeling in Biology

We have seen that Schwinger emphasized that an engineer's intermediate theory "looks in both directions" toward the fundamental theory and toward the experimental data. In this respect the engineering approach fits with recent proposals by systems biologists in dealing with multiscale phenomena in biology.[8] These proposals call for "middle-out" modeling strategies. As such, they privilege the mesoscale both spatially and temporally. Denis Noble and collaborators employ these middle-out strategies to explain and understand the behaviors of various biological phenomena. A particular focus of Noble's is on understanding the heart:

> [C]omplex systems like the heart are inevitably multiscalar, composed of elements of a diverse

---

[8]See a discussion of this in (Batterman and Green, 2020).

nature, constructed spatially in a hierarchical fashion. This requires linking together different types
of modeling at the various levels. It is neither
possible nor explanatory to attempt to model at
the organ and system levels in the same way as at
the molecular and cellular levels. To represent the
folding, within microseconds, of a single protein
using quantum mechanical calculations requires
months of computation . . . . Even if we could do it,
we would still need to abstract from the mountain
of computation some explanatory principles of
function at the cellular level. Furthermore, we
would be completely lost within that mountain
of data if we did not include the constraints
that the cell as a whole exerts on the behavior
of its molecules. This is the fundamental reason
for employing the middle-out approach. In mul-
tiscalar systems with feedback and feed forward
loops between the scale levels, there may be no
privileged level of causation.
    (Bassingthwaighte et al., 2009, p. 597)

Noble explains why this middle-out strategy is superior to a
bottom-up approach that starts with fundamental theory.

The reason is that this way we can select what we
are interested in. As we reach down towards lower
levels, we can rely on our higher-level analysis
to identify just those features of the lower-level
mechanisms that are relevant, and we ignore the
rest. The lower levels are seen through the filter of
the higher level. This allows us to highlight what is
important in the otherwise overwhelming mass of
data. It greatly reduces the amount of information
that we must carry over from one level of analysis
to the other. (Noble, 2006, p. 81)

I think the best way to understand claims like this is to see
them as recognizing that the upper-scale behaviors are to

a large extent independent of, or autonomous from, lower-level details. The overwhelming mass of data are irrelevant. The relevant mid-level details are represented by the variables and parameters that suppress much of the lower-scale data. In contexts involving many-body systems (one can consider the heart as one such system of heterogeneous composition), these variables and parameters are those that often reflect mesoscale structures and the transport properties of conserved quantities.

Noble also emphasizes that multiple models are needed because the equations used are scale-dependent:

> The central feature from the viewpoint of biological modelling can be appreciated by noting that the equations for structure and for the way in which elements move and interact in that structure in biology necessarily depend on the resolution at which it is represented. Unless we represent everything at the molecular level which . . . is impossible (and fortunately unnecessary as well), the differential equations should be scale-dependent. (Noble, 2011, pp. 61–62)

We have, in fact, seen the scale dependence of such equations repeatedly. The hydrodynamic methods outlined in Chapter 3 aim specifically to derive the mesoscale spin-diffusion equation (3.7). The Brownian equation (4.3) and the equation expressing stress–strain relations for a composite material (5.5) are also defined only at mesoscales.

6.4.1 Modeling Bone Fracture

Let us look at another example of a middle-out engineering approach.[9] Human bone is an active material, unlike steel. Bone structures change as we age, bones can heal after breaking, etc. Despite these important differences, there are

---

[9]This section is heavily influenced by joint work with Sara Green (Batterman and Green, 2020).

many similarities between the methods used to model material like steel and this approach to modeling the mechanics of bone.

At macro-scales, whole bone looks relatively homogeneous and dense, and one would expect continuum equations to be relevant in describing, say, the loading behavior of the human femur. If we break a bone we will see very complicated hierarchical and heterogeneous structures. As is evident in figure 6.2, there are a variety of structures at different spatial scales.[10] At the scale of 10^{-3}m there are two distinct tissue types—trabecular bone and cortical bone. The characteristics and distribution of these materials determine various physical/physiological properties of the bone at continuum scales (the scale of whole bone). They also allow for modeling of behaviors of medical relevance, such as fractures and osteoporosis.

Cortical bone comprises the dense hard shell of bone. It surrounds trabecular bone, which consists of trabeculae. These structures are more open and porous than the cortical bone. They are largely responsible for shock absorption. Together, at this mesoscale, they allow the continuum behavior of bone to be reasonably lightweight yet still fracture-resistant.

As shown in figure 6.2, if we were to zoom in further we would see new structures. From those lower-scale perspectives, the mesoscale structures of cortical and trabecular bone cannot be identified. This is similar to the inability of a lattice-scale perspective to allow one to characterize the mesoscale structures in a steel beam, for example. The mesoscale structures, recall, are described in terms of correlation functions and not individual atoms.

If we recall the discussion determining the stiffness tensor for a two-phase material in Chapter 5, we see that upscaling to determine the continuum elastic properties of bone presents a similar problem. In the earlier discussion we saw that the geometry and topology of the mesoscale RVEs were the most

---

[10]This is further evidence that what counts as mesoscale is relative to the particular problem at hand.

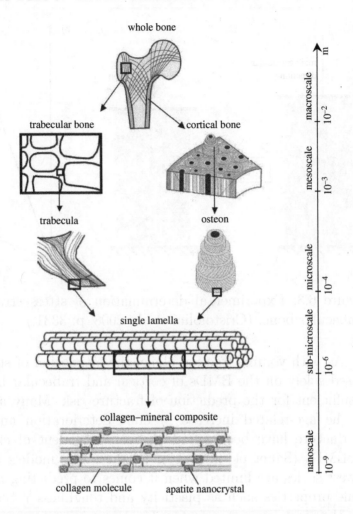

Figure 6.2: Bone structure and spatial scales. (Sabet et al., 2016, p. 2.)

important features. The current situation is largely the same, though as mentioned, the active nature of bone makes the problem even more difficult. One feature that is also important (analogous in a certain sense to the volume fraction ϕ in our earlier discussions) is that cortical bone and trabecular bone have different bone-material-densities (BMD). The different kinds of bone respond differently to loading, as is evident from experimental tests to determine their stress–strain curves. See figure 6.3.

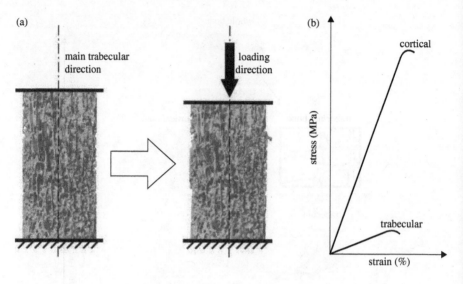

Figure 6.3: Experimental determination of stress–strain for trabecular bone. (Cristofolini et al., 2008, p. 3331.)

As with volume fraction alone, the estimation of stiffness based solely on the BMDs of cortical and trabecular bone is insufficient for the prediction of fracture risk. Many aspects of the age-related increase in bone deterioration and risk of fracture have been found to be independent of changes in BMDs (Sabet et al., 2016). Furthermore, models at the lowest scales are limited when it comes to predicting macro-scale properties such as plasticity and toughness.[11] So what are the most important factors for determining fracture risk? Unsurprisingly, they are geometric and topological features of bone at the mesoscale. Many fractures in elderly humans occur in the hip at the "neck" of the femur. As is evident in the whole bone representation at the macro-scale in figure 6.2, the neck is primarily composed of trabecular bone, with only a thin covering of cortical bone.

---

[11] *Stiffness*, as we have seen, reflects a material's resistance to deformation under loading. *Toughness* reflects a material's ability to deform plastically without fracture. A pane of glass is relatively stiff, but not so tough.

Without going into much detail here, the mesoscale geometric and topological differences between young males and elderly females can be striking. See figure 6.4. The more open geometry of the trabecular network in the elderly female makes fracture much more likely. If we were to zoom in to lower scales, we would not gain much relevant detail that would help in fracture prediction.[12]

Thus, as with the cases of inactive materials discussed in Chapter 5, we see that geometry and topology in the relevant RVEs are the most important mesoscale features for

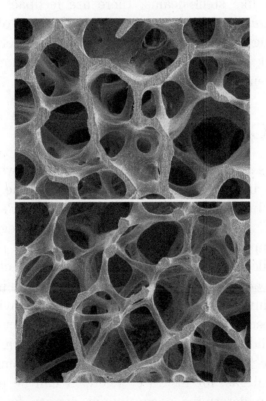

Figure 6.4: Spongy bone of a 21-year old male (top) and a 65-year old female (bottom). These images are used with permission from James C. Weaver. The scale is 3 mm × 3 mm for each. (Ritchie et al., 2009, p. 46.)

---

[12]For more details see (Batterman and Green, 2020).

determining large-scale behaviors. Structures at even lower scales are largely irrelevant.

This section has focused primarily on similarities of methods for complex systems, whether in condensed-matter physics, materials science and engineering, or biology and biomechanics. The main goal was to further stress the importance of mesoscale structures—a feature of Schwinger's engineering approach—and middle-out strategies for understanding the behavior of complex many-component systems. That said, I should note that even in the contexts of inactive materials like steel beams, there are feedback mechanisms in *computational modeling*, just as in the case of bone where remodeling at lower scales forces changes in upper-scale behaviors. Mark Wilson has stressed such computational strategies in his recent book, *Physics Avoidance* (Wilson, 2018).

6.5 Conclusion

We have seen that Batchelor distinguishes between upscaling problems in the "physics domain" and in the "engineering domain." In this chapter we discussed Schwinger's "engineering approach" to particle physics, and I have begun to argue for the superiority of approaching many-body systems from this middle-out, mesoscopic perspective. In effect, I think the *narrow* reading of "physics domain problems" is inappropriate for most investigations into the continuum behaviors of many-body systems. Rare exceptions, where such direct fundamental to non-fundamental connections are successful, are often cases where the fundamental microstructures are homogeneous, as in an ideal gas, for instance. There, it can be reasonable to upscale to determine a continuum density for such a gas. Simple volume averaging—counting the number of molecules N in a RVE, dividing by the volume V, and finally taking limits as $(N, V) \to \infty$—will yield an *effective density* at continuum scales. In general, however, we need to recognize that inter-theory relations of the form "theory X is more fundamental than theory Y" require mesoscale mediation rather than direct, straightforward upscaling.

What is more, we have seen, at least in the statistical mechanical context where the Fluctuation–Dissipation theorem holds, that mesoscale structures *will* exist (represented by order parameters and material parameters), and they provide the requisite mediating connections. Of course, it is almost certain that the upscaling that determines effective transport coefficients will *at best* determine well-defined *bounds* on the values for those coefficients. (As we saw in Chapter 5, one really requires an infinite number of correlation functions to uniquely determine the values.) Nevertheless, without the kind of geometric and topological information captured by the mesoscale, correlational parameters (these are Schwinger's "effective lumped constants"), the upscaling task will be hopeless for systems that are heterogeneous at lower scales.

In addition, we looked at an example from biology that also involves upscaling using mesoscale mediating connections. Denis Noble has, in fact, advocated the middle-out approach in biology for many years. The "engineering" approach is widespread in science, and yet it is widely ignored in the philosophy of science.

Let us take stock of the general problem once again using Kadanoff's metaphor of an inner world in the midst of an outside world. The Brownian particle was the "inner" world weakly exchanging conserved quantities with the "outside" world of the ambient fluid/solvent/matrix. Einstein, though, determined the effective viscosity for a third world, considered to be a continuum fluid. It is here that the talk of "worlds" gets a bit awkward. Instead, we should shift to talk of scales: the modeling of many-body systems involves three scales. The inner world is the (meso) scale of the RVE. The Brownian particles play the role of heterogeneous structures weakly interacting with a matrix representing the outer world which is the (micro) scale of the ambient fluid/solvent. And finally, there is the world of our ordinary experience of many-body systems. This is the (macro) world accurately represented using the continuum equations of fluid mechanics and of the continuum mechanics of material behaviors.

In the next chapter I examine a problem that confronts much of scientific theorizing. What are the right or *natural* variables with which to describe the world? I will argue that the discussions so far provide some compelling reasons to hold that mesoscale parameters are in many contexts much more natural for describing the behaviors of large systems than are the lower-scale variables and parameters that appear in the more "fundamental" theories of the constituents of those systems. The key is to understand what allows us to account for the relative autonomy of upper-scale less fundamental theories, from lower-scale more fundamental theories. This is to say that the degree of naturalness of a variable is proportional to the variable's ability to aid in accounting for this relative autonomy.

Chapter 7

The Right Variables and Natural Kinds

7.1 Introduction

In this chapter I argue that mesoscale parameters—order parameters and material parameters—are *natural* variables in the sense that they are the best variables with which to characterize certain dominant, *lawful* behaviors of many-body systems. Thus, this chapter aims to justify these variables/parameters as better able to figure in explanations, as better able to provide descriptions and understanding of certain behaviors, than supposedly more "fundamental" variables and parameters. The ultimate conclusion will be that a middle-out engineering approach to many-body systems is often ontologically superior to one based in fundamental theory. By ontological superiority I mean that these quantities or kinds allow for much more effective modeling of the mesoscale regularities exhibited by many-body systems. Furthermore, they also provide the best explanation for why continuum-scale regularities hold. In arguing this I am addressing metaphysical concerns about the proper way to carve nature at its joints. I will argue that, at least in the current context, we *do not* need to engage in metaphysical analysis.

A Middle Way: A Non-Fundamental Approach to Many-Body Physics. Robert W. Batterman, Oxford University Press. © Oxford University Press 2021.
DOI: 10.1093/oso/9780197568613.003.0007

In the next section I set up the problem by discussing James Woodward's take on variable choice in scientific contexts. He opts for a conception of naturalness (or a conception of a *right or correct* variable) that is ultimately tied to the aims or goals of scientific inquiry. And he contrasts this conception with a cluster of conceptions that come out of contemporary theorizing in metaphysics. These latter take natural properties to be fundamental and primitive.

My goal in what follows is to try to steer a middle path, arguing that we have theoretical/scientific reasons to consider some variables to be natural in an important but different, sense. Thus, I am in a way returning to debates about the nature of natural kinds that had a large presence in metaphysics and philosophy of science in the latter half of the twentieth century. Those debates focused on trying to analyze natural kinds in terms of their role in laws. I suggest that the philosophy of science debates about laws and natural kinds missed something important. There can be scientific reasons for treating some mesoscale variables as right or natural. Specifically, I will argue that the Fluctuation–Dissipation theorem *requires* that we theorize about many-body systems using parameters or variables that represent mesoscale structures of those systems.

7.2 Woodward on Variable Choice

In "The Problem of Variable Choice" (2016), James Woodward raises some questions about finding the correct variables that can figure in causal analysis and causal explanation. He distinguishes the problems of interest from philosophical presentations of properties that "seem 'unnatural' or 'gerrymandered' or even 'pathological', with 'grue' perhaps the most prominent example" (Woodward, 2016, p. 1048). For Woodward, finding the right causal variables is a genuine problem in many areas of science:

> Indeed, it is a very common worry in many scientific disciplines (including neurobiology,

> cognitive psychology, psychiatry, genetics, and macoreconomics) that theorizing at present is conducted in terms of the wrong variables and this seriously impedes theory construction, including causal analysis. (Woodward, 2016, p. 1049)

In the context of providing a justification for the relative autonomy of continuum-scale physics from fundamental atomic or subatomic details, we have a similar problem. It just does not seem likely that *in principle*, purely bottom-up derivations from an unattained (and perhaps unattainable) completed physics will give us an account of that autonomy. (Recall the discussion in Chapter 2.)

Woodward notes that in complex systems like the brain and the weather, it is best to approach theoretical (causal) understanding by aggregating the myriad of variables that are used to describe the many components of those systems. That is, we often must coarse-grain or "parameterize" and thereby reduce the number of degrees of freedom that such complex systems possess. While I think it is true that reducing the number of degrees of freedom is often a good move from a pragmatic or computational perspective, I want to argue that there are non-pragmatic reasons for doing so as well.[1] In other words, I want to claim that there are theoretical reasons that allow for the determination of the right kind of variable reduction and the right kind of parameterization. I take these reasons to motivate an understanding of the bulk behavior of many-body systems in terms of correlational structures that exist at mesoscales.

For Woodward, two main goals of scientific inquiry are causal representation and causal explanation. Furthermore, on his view, the proper variables are those that lead to representation and explanation in terms of manipulation and control. In fact, in the context of many-body systems (one

---

[1]One should also remember that the right kind of aggregating, in most instances, involves homogenization/upscaling strategies that are much more complicated than simple averaging. See section 5.2.

of his examples is climate modeling), he argues that causal representation and causal explanation should aim for aggregative variables that lead to models employing generalizations that are more, rather than less, stable in the sense of holding true in a "wide range of background circumstances." Such variables, because they lead to more stable generalizations, will provide "more information relevant to manipulation and control" (Woodward, 2016, p. 1053).

Therefore, Woodward's approach to variable or parameter selection is normatively tied to the goals of causal explanation and causal analysis. Potential criteria for selecting the right variables are to be evaluated by their success in promoting such aims. By tying the criteria to these norms, Woodward is giving up on *direct attempts* to argue that certain variables or parameters are *natural simpliciter*. In other words, a variable is right on Woodward's view if it promotes the activities of causal representation and causal explanation as understood in an interventionist framework (Woodward, 2005).

As a result, he distances himself from more recent metaphysical attempts to identify the right variables as natural kinds, or perfectly natural properties. Often, he notes, according to various metaphysical views these natural properties or kinds are to be provided by an examination of the properties that appear in a fully *fundamental* physics (Woodward, 2016, p. 1056). Sider endorses one version of such a view:

> The heart of metaphysics is the question: what is the world ultimately, or fundamentally, like? And fundamentality is a matter of structure: the fundamental facts are those cast in terms that carve at the joints.
>
> The truly central question of metaphysics is that of what is *most* fundamental ... [W]e must ask which notions carve *perfectly* at the joints.
>
> (Sider, 2011, p. 5)

The philosophical literature on laws and natural kinds during the second half of the twentieth century largely attempted to find necessary and sufficient conditions for a generalization to be a law (Hempel, 1965; Goodman, 1983). Laws were supposed to relate natural kinds to natural kinds. These debates led to various attempts to specify what made a property natural. As Woodward notes, the predicate "grue" famously could not refer to a natural property in the way that "green" may very well so refer. It is fair to say that most of these discussions went round in circles, analyzing naturalness in terms of lawlikeness and lawlikeness in terms of naturalness.[2] More recent metaphysical discussions, including Sider's, have shifted away from the earlier more law-focused suggestions toward the idea that the natural variables are those appearing in theories that are *most fundamental* or most joint-carving. The analyses of the last century, according to these recent discussions, just did not pan out.

In hitching his notion of a good, or right variable to a normative goal of science, Woodward resists the more recent metaphysical turn. He summarizes this as follows:

> Thus, I propose replacing the role assigned to natural properties by metaphysicians with a methodological investigation into which variables best

---

[2]Fodor, in "Special Sciences," expressed the state of the debate as follows:

> If I knew what a law is, and if I believed that scientific theories consist just of bodies of laws, then I could say that P is a natural kind predicate relative to S iff S contains proper laws of the form $P_x \to \alpha_x$ or $\alpha_x \to P_x$; roughly, the natural kind predicates of a science are the ones whose terms are the bound variables in its proper laws. I am inclined to say this even in my present state of ignorance, accepting the consequence that it makes the murky notion of a natural kind viciously dependent on the equally murky notions of *law* and *theory*. There is no firm footing here. If we disagree about what is a natural kind, we will probably also disagree about what is a law, and for the same reasons. I don't know how to break out of this circle, but I think that there are interesting things to say about which circle we are in. (Fodor, 1974, p. 102)

serve various goals of inquiry. In this sense, [my discussion] may be viewed as an instance of a more general project of "replacing metaphysics with methodology." (Woodward, 2016, p. 1057)

Here Woodward cites Hitchcock's (2012) "Events and Times: A Case Study in Means–Ends Metaphysics" as a particular instance of this general project.

I am sympathetic to this proposal. However, it seems to me that there may be a way forward that endorses neither the contemporary metaphysical discussions of what is natural in terms of the *fundamental*, nor a full-fledged normatively relativized notion of what is natural. This is to say, that sometimes we have non-metaphysical and non-methodological resources with which to identify or delimit a privileged class of natural-kind variables. In fact, the context of a hydrodynamic/correlation function approach to many-body systems provides a compelling set of examples.

7.3 The Right (Mesoscale) Variables

The discussion in previous chapters has highlighted the role that RVEs play in determining effective, continuum material behaviors of various systems. In particular, following Kadanoff and Martin, as well as Batchelor, we have focused on bulk transport properties of many-body, heterogeneous systems. Our examples have included the hydrodynamic description of spin diffusion, Brownian motion, and various "static" problems from materials science. One important aim has been to provide our continuum theories with effective transport coefficients that fictitious homogeneous systems may possess, thereby allowing them to mimic the behavior of real heterogeneous systems of interest.[3]

---

[3]As we have seen, in doing this we are able to argue for the relative autonomy of the continuum theories.

In this section I would like to elicit a conclusion from the connection between disturbed equilibrium systems and statistical fluctuations emphasized by the Fluctuation–Dissipation theorem: the material and order parameters that code for correlational details present in the relevant RVEs are genuinely *natural* parameters or natural kinds. They are the right variables with which to model and investigate various aspects of the bulk behavior of many-body systems.

In order to do this, I suggest we focus on why some models, like the Ising model, have wide-ranging applicability. This discussion, therefore, contributes to a current, reasonably heated, philosophical debate about what makes such special or "minimal" models valuable.[4] Once we understand why those models are as fruitful as they in fact are, we will have a motivation for treating material parameters and order parameters as natural kinds in a non-pragmatic and non-normative sense. Also, such parameters will not be "natural" according to much of contemporary metaphysics where than concept is tied, rather closely, to "fundamentality."

In a nutshell, the argument is the following. The ability of such minimal/special models to represent correlations in RVEs explains the models' wide ranging applicability for representing regularities in bulk behaviors. Those correlations are, therefore, natural variables or parameters for characterizing continuum properties of many-body systems. As we have seen, these parameters are relatively independent of the fundamental variables describing the system at the lowest scales.

In an article entitled "Condensed Matter Physics: Does Quantum Mechanics Matter?" (1988), Michael Fisher argues that for the most part the title question must receive a negative answer: quantum mechanics *almost* does not matter for condensed-matter physics. Of course, condensed-matter physics, in the terminology of this book, is *relatively* (not fully) autonomous from quantum mechanics: we would not

---

[4]See (Batterman and Rice, 2014; Lange, 2015; Reutlinger et al., 2018, and references therein) for some of the details.

have stable states of matter without it.[5] Furthermore, just to be completely clear, neither Fisher nor I am claiming that fundamental physics is explanatorily worthless. There are many why-questions that require full-on appeal to the lower/lowest-scale details. It is just that, for a number of questions about bulk behavior of many-body systems, those more fundamental details are almost completely irrelevant.

Fisher notes:

> The basic problem which underlies the subject [condensed matter physics] is to understand the many, varied manifestations of ordinary matter in its condensed states and to elucidate the ways in which the properties of the "units" affect the overall, many-variable systems. In that enterprise it has become increasingly clear that an important role is and, indeed, should be played by various special "models." (Fisher, 1988, pp. 66–67)

Fisher holds that these models, like the Ising model, are special because, for example, they allow for an understanding of the universal behavior of dilute gases. And they transparently demonstrate that such behavior is

> in most regards quite independent of the chemical constitution of the molecules or atoms. One should not, therefore, be so surprised that other theoretical models that have been abstracted from complicated physical systems, like the basic Ising and Heisenberg models originally devised for magnetism, take on a life of their own quite regardless of the underlying atomic and molecular physics (for which quantum mechanics does matter). (Fisher, 1988, p. 67)

---

[5]Recall the discussion of autonomy in Chapter 2.

So for Fisher, the fact of universality explains the independent lives of such minimal models.[6] In other words, we can explain the utility of minimal models by appealing to the fact that such models are in the same universality class as the actual system of interest. Recall the discussion of section 2.2.3.

But, universal behavior *itself* requires an explanation. (See Chapter 2 as well as (Batterman, 2002, 2017, 2019; Batterman and Rice, 2014).) A consequence of such an explanation will also explain why minimal models like the Ising model are so useful and provide such insight into the nature of non-simple many-body systems (Batterman and Rice, 2014). The argument of section 2.2.3 shows that the explanation involves demonstrating that models like the Ising model are in the basin of attraction of the appropriate RG fixed point.

Nevertheless, those with reductionist leanings might hold that fundamental theory should explain this universality and the fecundity of minimal models. Fisher pushes back against this suggestion, and it is worth considering how, in light of our discussions of the Fluctuation–Dissipation theorem, we can provide a response to the reductionist. Put differently, can we give a non-RG, mesoscale first, argument for the role of minimal models in condensed-matter/many-body physics? The answer is "yes." Here is Fisher once again:

> ... the mere fact that such an 'artificial,' 'non-fundamental' model as the Ising model provides insight into a wide range of *contrasting* examples of condensed matter as anisotropic ferromagnets, gas-liquid condensation, binary alloys, structural phase transitions, etc., shows that the question of "connecting the models with fundamental principles" is *not* a very relevant issue or central enterprise. (Fisher, 1988, p. 67)

---

[6]This fits with one argument of Kadanoff's that I have discussed in (Batterman, 2017, pp. 568–569), where universality functions as an explanans.

In this quote, Fisher stresses the applicability of minimal models such as the Ising model across widely different systems and problems. From the point of view of mesoscale modeling using RVEs, we have seen that upscaling to determine effective material parameters such as viscosity and stiffness tensors requires information primarily about spatial and temporal correlations between structures in the RVEs. In flowing contexts, these are densities and gradients of densities of conserved quantities. In static cases they are shapes and topological features of inclusions of some materials in a matrix of another. In all cases, as we have seen, correlation functions are used to characterize these mesoscale features.

Consider the Ising model in figure 7.1. We can let the up and down arrows stand for spins in the context of ferromagnetism, for liquid and steam in a boiling kettle, for different alloys, or for atoms of different materials in a composite. The model allows for simple representations of correlations between such lower-scale components of many-body systems as the groupings demonstrate.[7] If our goal is to understand

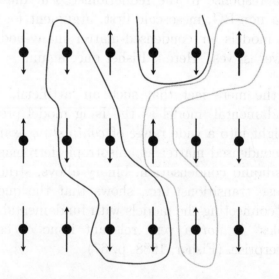

Figure 7.1: Minimal Ising model and RVE.

---

[7]Here we are circling back to the discussion in sections 1.4 and 1.5.

the continuum behaviors of many-body systems, given some knowledge of the components, models like this apply across a wide range of different cases.

So why are minimal models so apt? Why are they so widely applicable? The answer is because they do not model "fundamental" properties of systems, but they do model the *natural* properties of many-body systems: they allow for the *direct* coding of correlations at the scale of the RVEs. Those correlations, both in space and time, are all that matters for characterizing the bulk behavior of many-body systems.[8] The correlations are just what determine the values for the order parameters and material parameters that provide the material specificity in the bulk, continuum equations. In other words, they let us determine the constitutive equations, *à la* Hooke's law; namely, those relations defining the effective parameters for viscosity, stiffness tensors, spin-diffusion, etc. These constitutive equations, as we have seen, are required to solve the hydrodynamic equations for actual many-body systems.[9]

I have been arguing that the structures in the RVEs, for a wide range of heterogeneous many-body systems across different sciences and in different contexts, are the key ingredients in upscaling arguments. Ising models and other minimal models are designed to represent those essential mesoscale features that, given the Fluctuation–Dissipation theorem, we know must be present. It is this fact that justifies taking the mesoscale parameters as *natural* with respect to the bulk behavior of many-body systems.

---

[8]Another way of putting this, one which fits nicely with causal modeling strategies, is to say that these variables are the *difference makers* for the bulk behavior of many-body systems.

[9]Of course, after the discussion in Chapter 5, "hydrodynamic equations" generalizes to include various static materials science/engineering problems discussed in Batchelor's classification (Batchelor, 1974).

7.4 Another Minimal Model Example

In this section I would like to talk about another minimal model—one that has been used to understand various gross features of fluid flow at scales significantly above the molecular. This is the model of lattice gas automata discussed by Goldenfeld and Kadanoff (1999) and by Batterman and Rice (2014). While the systems modeled are not near-equilibrium many-body systems for which the Fluctuation–Dissipation theorem holds, the model is successful because it focuses on (represents) the most important feature at the mesoscale; namely, gradients of conserved quantities such as momentum.

7.4.1 The Model: Lattice Gas Automaton

The aim of the minimal model, called a Lattice Gas Automaton (LGA), is to understand the large-scale patterns witnessed in fluid flow. This remarkably simple model does, in fact, recover a host of features observed in real fluids at these large (continuum) scales.

Consider a set of point particles confined to move on a hexagonal lattice. Each particle can move in one of six directions, so we attach a vector to each particle. (See figure 7.2.) The rules are as follows. Between (1) and (2) the particles move in the direction of their arrow to their nearest-neighbor site. Then, if the momentum at that site sums to zero, the particles undergo a collision resulting in a jump of 60°, as shown in (3).

Now take many many particles, many many iterations of this update algorithm, and perform some coarse-grained average to yield macroscopic fields like number and momentum densities. The result will accurately reproduce many of the macroscopic features of fluid flow. In particular, it has been shown that this model reproduces quite accurately the parabolic profile of momentum density that is characteristic of incompressible laminar flow through a pipe (Kadanoff et al.,

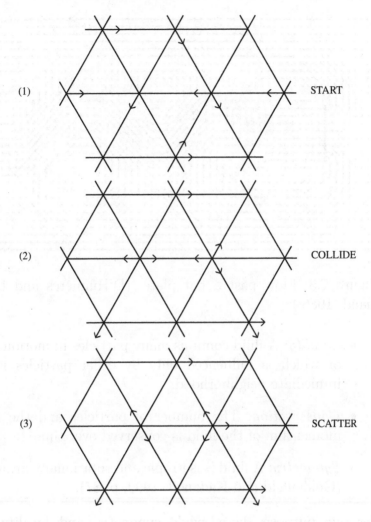

Figure 7.2: Update algorithm for lattice gas. (After (Golden-feld and Kadanoff, 1999).)

1989). The image in figure 7.3 also provides some idea of how well this model reproduces qualitative features of fluid flow past obstacles.

Goldenfeld and Kadanoff note that the Navier–Stokes equations show how the velocity of a fluid at one point in space affects the velocity of the fluid at other points in space. It is this fact together with the following fundamental features of the Navier–Stokes equations that motivate the LGA model:

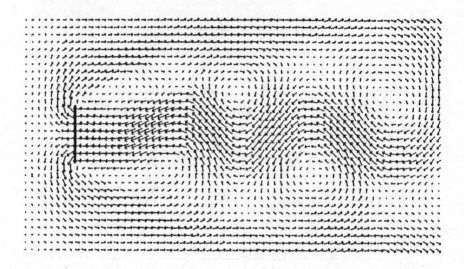

Figure 7.3: Flow past a flat plate. (D'Humières and Lalle-mand, 1986.)

- *Locality*: A fluid contains many particles in motion each of which is influenced only by other particles in its immediate neighborhood.

- *Conservation*: The number of particles and the total momentum of the fluid is conserved over time.

- *Symmetry*: A fluid is isotropic and rotationally invariant (Goldenfeld and Kadanoff, 1999, p. 87).

For our purposes here, while symmetry and locality are important, it is the conservation of particle number and momentum that is crucial. The LGA is designed to display transparently the gradients of the number of particles (their densities) of the fluid and the changes in its momentum (the velocities). In other words, correlations among particles and their evolutions over time are tracked by the LGA. The model itself looks *nothing like* the continuum fluid, *nor does it look anything like* the "fundamental" collection of molecules that constitute it. (Of course, the same is true for the Ising model.) Nevertheless, for a wide range of actual fluids, this model provides an exceptionally good scheme with which to compute the evolution of fluids at large scales. A "fundamental" model that actually aims to track the trajectories of the individual molecules would be woefully inadequate. The *natural* variables or parameters simply are not discernible at the molecular, "fundamental," scale.

The LGA model perspicuously represents mesoscale structures in fluid flow for a wide range of fluids. It allows for a simple representation of the essential elements of fluid RVEs in situations where the (many-body) fluids are far from equilibrium. In our previous discussions we have mainly focused on near-equilibrium systems for which the Fluctuation–Dissipation theorem holds. However, the lessons generalize. In the context of understanding the continuum behaviors of many-body systems in general, mesoscale variables—reflecting correlations among components and gradients of conserved quantities—are the natural variables for characterizing bulk behaviors whether equilibrium or non-equilibrium. Furthermore, these variables provide essential mediating connections between relatively autonomous continuum-scale theories and models and "fundamental" lower-scale models and theories.

7.5 Conclusion

It is a fact that minimal models such as the Ising model and the Lattice Gas Automaton are widely applicable. That applicability should be explainable. The explanation is related to the RG story that was told in Chapter 2 where appeal was made to a stability transformation that perturbs an actual fluid's Hamiltonian into, say, the Ising Hamiltonian in such a way that the continuum behavior of the fluid remains unaffected.

Another way to understand this stability is to consider the consequences of the Fluctuation–Dissipation theorem for the representation of the bulk behavior of large many-body systems. Fluctuations from equilibrium are described by correlation functions in space and time. The strength of those correlations is related to the dissipation strength—the frictional force operative in the dynamics. The fact that correlations are essential to describing the fluctuations explains why models that can represent correlated behavior (and its evolution) are themselves so useful. The minimal models are, either intentionally or by happy accident, well suited to represent the mesoscale structures that the theorem guarantees must be present. Initially, I think, the successes and wide applicability of models like the Ising model were an unexpected bonus. Models like the LGA built upon a theoretical understanding of what actually makes minimal models so successful. That understanding was, I believe, an implicit appeal to those models' utility in allowing for mesoscale representations of correlational structures in relevant RVEs.

That all many-body systems (in the linear near-equilibrium regime defining the hydrodynamic description) obey the Fluctuation–Dissipation theorem tells us that mesoscale correlational structures will be present. Furthermore, these are just the features that are required to describe the bulk transport of the important, measurable, conserved quantities. We therefore have scientific, theoretical reasons to take the parameters that describe that behavior to be natural, uncontrived, and non-gerrymandered. They are the right variables to describe many-body systems in this regime and at large/continuum scales. Furthermore, we neither have to appeal to fundamentality to characterize the natural "joint-carving" parameters or variables, nor must we relativize our conception of naturalness to any particular epistemic goals.

By focusing on the mesoscale—a midway between the atomic and the continuum—we have found a middle way between a fully metaphysical approach and a fully, normatively relativized, methodological approach to when a variable can be a natural kind.

Chapter 8

Conclusions

8.1 Foundational Problems vs. Methodology

The preceding chapters aimed to shine a spotlight on a ubiquitous methodology in physics, materials science, and biology for studying the bulk behavior of many-body systems. I have labelled this set of methods "hydrodynamic methods" following the lead of Kadanoff and Martin (1963). I believe that reflection on this methodology is almost, if not completely, absent in the philosophy of science literature. This is a missed opportunity.

A (the?) reason for failing to notice the importance of this widespread, widely applicable scientific methodology is, I contend, a rather myopic focus on important philosophical problems of a *foundational* nature. Philosophers of physics have, I would claim, primarily been trained to focus on foundational problems in various fields. We apply our analytical skills with an aim toward resolving conceptual problems that appear in many different physical theories. I think this is a legitimate and important contribution to science and to the philosophy of science. It has its genesis, I believe, in the development of logical positivism and logical empiricism. Of course, this is not the place to engage in the history and sociology of contemporary philosophy of physics. But as I suggested in

A Middle Way: A Non-Fundamental Approach to Many-Body Physics. Robert W. Batterman, Oxford University Press. © Oxford University Press 2021.
DOI: 10.1093/oso/9780197568613.003.0008

sections 1.1 and 1.2, concerns about fundamentality in the philosophy of physics are also often addressed through a search for proper foundations—foundations that aim to secure the fundamental theory to a firm logical or axiomatic bedrock. And once this has been done, it is asserted (explicitly or implicitly) that we have reason to believe that, *at least in principle*, any connections between fundamental theory and less fundamental theories are *merely* a matter of mathematical derivation. The only impediment to accounting for a hierarchy of theories is pragmatic.

To the contrary, I have argued that a proper understanding of the relation "theory X is more fundamental than theory Y" is typically much more difficult than this "in principle" claim[1] suggests. In fact, most successful upscaling relations between theories in this hierarchy are mediated by a mesoscale first approach. Direct derivational connections between, say, continuum theories of many-body systems and fundamental theories of their atomic constituents, are rare, and depend upon the special nature of the systems. Specifically, direct connections depend upon the systems being quite homogeneous in their lower-scale makeup.

The hydrodynamic methods exploit the fact that there *must be* heterogeneous structures at scales intermediate between fundamental atomic levels and continuum descriptions. These mesoscale, heterogeneous structures can be used to describe and explain non-equilibrium behaviors of many-body systems. Specifically, the focus is on understanding transport—currents that result from spatial and temporal non-uniformities with respect to conserved quantities. For example, if there is a temperature gradient, there will be heat flow; if we have an electrical conductor in an electric field, there will be a current proportional to the strength of the field. In all such cases, there are material variables or parameters that characterize these behaviors.

---

[1] It is indeed a claim, and not an argument.

Statistical mechanics, considered as a bottom-up approach based on fundamental microstructure and the dynamics of that structure (a kinetic theory approach), is not remotely well suited for determining values for these material parameters or transport coefficients. But the hydrodynamic approach to non-equilibrium, based upon a linear response to external pushes or internal fluctuations away from equilibrium, has been remarkably successful. These successes and the philosophical lessons for questions about upscaling, reduction, and inter-theory relations in general have been the focus of the preceding chapters.

An immediate philosophical lesson is that the hydrodynamic methodology allows for an explanation of the relative autonomy of upper-scale continuum theories from lower-scale, more fundamental theories. It is a remarkable fact that theories of continuum mechanics and fluid dynamics survived the atomic revolution. After all, as I have noted many times, these theories are ontologically incorrect. They posit no sub-continuum structure yet they are spectacularly successful in their applications to real-world engineering problems. Reductionistic attempts to understand this success have, as I have argued in Chapter 2, fallen short of the mark.

8.2 Autonomy and Heterogeneity

Chapter 2 connected this explanatory challenge to the fact that many upper-scale patterns of behavior are multiply realized. Equivalently, in physics terms, these patterns are universal. Here I just want to reemphasize the connection between explaining the relative autonomy of continuum theories from atomic/discrete, more fundamental theories and the problem of multiple realizability. I raised the issue of relative autonomy by posing the following question:

- (**AUT**) How can systems that are heterogeneous at some (typically) micro-scale exhibit the same pattern of behavior at the macro-scale?

The assertion that systems differing in their microstructure nevertheless exhibit the same patterns of upper-scale behavior is a claim about multiple realizability or universality. So, I am claiming that multiple realizability and relative autonomy are intimately related. The Navier–Stokes and Navier–Cauchy equations capture patterns of behavior at continuum scales. For example, laminar flow in a pipe (for various kinds of fluids) exhibits a parabolic velocity distribution despite the differences in the molecular makeup of the fluids. So, that pattern of parabolic velocity distribution is multiply realized. If we can explain how such common behavior is possible, despite those *fundamental* differences, then we have an answer to (**AUT**).

Such an explanation can proceed in a couple of ways. If we can show that the common macroscopic behavior is stable under the perturbation of the molecular details of one fluid into those of another, then we have shown that the those molecular details are essentially irrelevant for that macroscopic behavior. This is what RG-like arguments can provide. See the discussion of the λ-transformation in section 2.2.3.

An alternative way to think about answering (**AUT**) is to consider the question from the point of view of the theory of homogenization. From this point of view the goal is to show, for an individual system that is heterogeneous at some lower scale, that we can find an equivalent homogeneous system— one that displays the same behavior at continuum scales. In other words, can we find a system that displays no structure at any scale, but that also exhibits the same behavior as the actual heterogeneous system at hand? If we can do this, then we know that the upper-scale behavior can be characterized without referencing any actual lower-scale details.

Here too, we will have shown that molecular/atomic details are irrelevant for the upper-scale behavior. And, in so doing, we have shown how the equations governing the continuum behavior can still apply despite the actual discrete nature of matter. Of course, the means by which we can demonstrate the existence of an equivalent homogeneous system is by determining an *effective* value for the relevant material parameter occurring in the continuum governing equations. I believe that while the RG exploits the self-similarity of systems near criticality, and homogenization does not need to, the overall aims of the two mathematical schemes are quite similar. The goal is to extract a phenomenological (universal) description at continuum scales from systems that genuinely differ—that are heterogenous—at lower scales.

It appears that there are two senses of "heterogeneity" at play here as well. In the context of explaining how multiple realizability or universality is possible, we are interested in distinct systems that differ in their lower-scale structures. Thus, the sense here is that heterogeneous *systems* display the same continuum patterns of behavior. RG arguments show that the details that genuinely distinguish the systems are by and large irrelevant for that behavior. On the other hand, in the context of upscaling via homogenization, we look at a single system that exhibits heterogenous (non-uniform) structure at a mesoscale. We then employ correlation functions describing the heterogeneous mesoscale structures to determine effective material parameters that figure in the continuum equations.

In the RG, multiple systems scenario, showing that the lower-scale differences between the systems are irrelevant, also demonstrates the extent to which the details of any individual system are irrelevant for its behavior at criticality—for its continuum behavior. Thus, the two senses of heterogeneity and the two ways of answering (**AUT**) are intimately related.

8.3 Brownian Motion and the F–D Theorem

The upscaling, mesoscale homogenization methodology had its genesis in Einstein's work on Brownian motion. In fact, as I note in Chapter 4, Einstein also paved the way for the hydrodynamic methods by providing the first expression of the Fluctuation–Dissipation theorem.

He connected the random zig-zag motion of the Brownian particles to the diffusion of the particles in their surrounding solvent. Here I provide another quick argument for the Fluctuation–Dissipation theorem due to Kubo (1986). Einstein's work led to the following expression for the diffusion constant D for Brownian Motion:

$$D = \mu k_B T \tag{8.1}$$

where μ is the mobility of the Brownian particles (it is inversely proportional to the viscosity, η, of the surrounding fluid), k_B is Boltzmann's constant, and T is the absolute temperature. The diffusion constant is defined as follows:

$$D = \lim_{t \to \infty} \frac{\langle \Delta x(t)^2 \rangle}{2t} \tag{8.2}$$

with $\Delta x(t) = x(t) - x(0)$ being the displacement of a Brownian particle in one dimension (say, the x-axis) after time t. This displacement exhibits a Gaussian distribution with variance $2Dt$. This can be related to the mean free path length l (the average distance between collisions), the mean free time τ (the average time between collisions), and the velocity v, as follows:

$$D = \frac{l^2}{2\tau} = \langle v^2 \rangle \tau \tag{8.3}$$

Using the equipartition theorem

$$m \langle v^2 \rangle = k_B T$$

with m the mass of the Brownian particle we get

$$D = k_B T \frac{\tau}{m} \tag{8.4}$$

Comparing this result with Einstein's equation (8.1), we have the following expression for the mobility of the Brownian particles:

$$\mu = \frac{\tau}{m} \tag{8.5}$$

Kubo (1986, p. 331) notes that

> [t]his derivation is different from [Einstein's] original one but it is instructive. Diffusion is a direct consequence of fluctuations of the velocity of the Brownian particle... Thus, fluctuation and dissipation are two aspects of a single phenomenon and thus are necessarily related to each other. This is the general concept of the FD theorem, of which Einstein's relation [equation (8.1)] was the first example.[2]

Finally, if we note that the displacement of the Brownian particle $\Delta x(t)$ can be related to its velocity change

$$\Delta x(t) = \int_0^t v(t')dt'$$

we get the explicit connection between the mobility of the particles and a correlation function in velocities:

$$\mu = \frac{1}{k_B T} \int_0^\infty \langle v(0)v(t) \rangle dt \tag{8.6}$$

This is a simple statement of the Fluctuation–Dissipation theorem in the context of Einstein's work on Brownian motion. Compare equation (8.6) to equation (4.20).

This theorem justifies the use of equilibrium statistical mechanics to treat non-equilibrium physical processes in terms of correlation functions. That is to say, it is responsible for the very possibility of the hydrodynamic methods for

---

[2]Kubo (1986, p. 331) actually allows for an external force such as gravity to play a role here. The only difference is that there will be drift in the direction of the gravitational gradient. However, the general expression of the Fluctuation–Dissipation theorem is the important consequence here.

studying the bulk transport behaviors of many-body systems. It is remarkable that this hydrodynamic methodology, in the context of Brownian motion and fluid flow, extends rather naturally to what at first seems to be a set of very different problems, including the bending of beams and the conduction of heat in a copper plate. In this latter context, we have seen in Chapter 5 that an homogenization methodology plays an essential role. In fact, both methodologies are intimately related, as Einstein's discussion of Brownian motion demonstrates. Both methodologies privilege the mesoscale, and both involve describing structures at the mesoscale in terms of correlation functions.

8.4 A Middle-Out/Engineering Methodology

The successes and the ubiquity of these two methodologies suggest that a mesoscale first, middle-out methodology is preferable to a reductionist, bottom-up approach to understanding bulk behaviors of many-body systems. In Chapter 6 I argued for this, building upon Julian Schwinger's "engineering approach" to particle theory. Schwinger argued that the "true role of fundamental theory is not to confront the raw data, but to explain the relatively few parameters of the phenomenological theory in terms of which the great mass of raw data has been organized" (Schwinger, 1969, p. 19). Schwinger's "source theory" was an attempt to realize this middle-out approach for strong interaction phenomena, and was really a precursor to contemporary effective approaches to quantum field theory.

The main argument of Chapter 6 was to show that mesoscale parameters or properties, specifically order parameters and material parameters, defined in terms of correlation functions, are required for making the connections between fundamental theories and experimental data determined at continuum scales. They mediate or intervene at scales between the atomic and the macroscopic.

The argument reflects the conclusions of earlier chapters that direct connections between fundamental theories and less fundamental, upper-scale, theories are exceedingly rare. It is fortunate for us that the world allows for a layered structure where correlational structures of aggregates of fundamental components of many-body systems can exist. I doubt very much that science as we know it would be possible in a world where scale separation was not possible.[3] Recall the speculations of section 2.4.

In fact, the Fluctuation–Dissipation theorem provides a guarantee that there will be such correlational structures that can be represented by order parameters and the like. I showed how the net magnetization (an order parameter) for a ferromagnet captures the ordered state and the breaking of rotational symmetry as a function of temperature. In the context of statistical mechanics, such an order parameter represents correlations among atomic-scale spins. Therefore, it really is coding for structures (densities and gradients of densities) at mesoscales. These are just the structures that define the hydrodynamic modes—long-lasting, measurable features of many-body systems in near- but out-of-equilibrium states.

8.5 A Physical Argument for the Right Variables

Finally, in Chapter 7 I examined whether the previous discussions could lead to an analysis of the nature of natural kinds. I argued that the order parameters and material parameters should be taken to be the natural or the *right* variables for characterizing the bulk behavior of many-body systems. The literature on natural kinds in the last century focused largely on relating naturalness to lawfulness and to distinguishing bad predicates (grue) from good predicates (green). This literature

---

[3]It is clear that we would not exist, as evolution needs to key on relatively stable regularities (patterns) at human scales.

also connected natural-kind terms to those that can figure in inductive inference, and to questions about counterfactual support.

I think it is fair to say that contemporary metaphysicians have moved on from much of these debates to discussions that relate naturalness to fundamentality. I also think it is fair to say that philosophers of science have largely abandoned the quest for an analysis of natural kinds in relation to lawlikeness. Despite this, I think there are theoretical/scientific reasons for treating some mesoscale variables as the natural or the right variables for scientific theorizing. So, rather than looking to philosophical intuition about laws or to metaphysical research into the nature of fundamentality, I think certain aspects of the hydrodynamic methodology can provide a means for identifying at least some natural kinds or variables.

Thus, there are *physical* reasons for focusing on the geometric and topological features in mesoscale RVEs. These structures, at least in the cases where we are dealing with a hydrodynamic description, are guaranteed existence by the Fluctuation–Dissipation theorem. Consideration of these structures are necessary for the determining the bulk transport properties of large systems that are heterogeneous at lower scales. They are also the ingredients needed to explain the successes and the robustness of continuum descriptions of many-body systems.

I have been arguing that we philosophers of science/physics have failed to pay attention to a ubiquitous and important scientific methodology. In so doing, our attention has primarily focused on what I call foundational problems in physical theories. These are indeed important. But, in focusing only on, say, bottom-up Boltzmann equation approaches to non-equilibrium statistical mechanics, we have almost completely failed to notice the middle-out strategy to non-equilibrium theory that condensed-matter physicists actually use and from which we can get reasonable values for material parameters and order parameters.

Of course, there are lessons here for more general debates about reduction, emergence, and hierarchies of theories. I have tried to argue that direct reductionist connections between more fundamental theories and less fundamental theories are very few and far between. However, I have also argued that full-on emergence characterized by the complete autonomy of the emergent theory from the more fundamental theory is also mistaken.[4] The upper-scale, less fundamental theories enjoy relative autonomy (or a kind of conditional independence) from their more fundamental partners. The key to understanding the nature of this autonomy is the same as that which allows mediated non-reductionist connections between the partners: namely, the mesoscale parameters that characterize the important/dominant features of the relevant RVEs.

---

[4]At least it is, in the context of condensed matter and materials science. I make no claims about mind–brain relationships.

Bibliography

Maedeh Amirmaleki, Javad Samei, Daniel E. Green, Isadora van Riemsdijk, and Lorna Stewart (2016). 3d micromechanical modeling of dual phase steels using the representative volume element method. *Mechanics of Materials*, 101: 27–39.

James Bassingthwaighte, Peter Hunter, and Denis Noble (2009). The cardiac physiome: Perspectives for the future. *Experimental Physiology*, 94 (5): 597–605.

G. K. Batchelor (1974). Transport properties of two-phase materials with random structure. *Annual Review of Fluid Mechanics*, 6: 227–255.

Robert W. Batterman (1998). Why equilibrium statistical mechanics works: Universality and the renormalization group. *Philosophy of Science*, 65: 183–208.

Robert W. Batterman (2000). Multiple realizability and universality. *The British Journal for the Philosophy of Science*, 51: 115–145.

Robert W. Batterman (2002). *The Devil in the Details: Asymptotic Reasoning in Explanation, Reduction, and Emergence*. Oxford Studies in Philosophy of Science. Oxford University Press.

Robert W. Batterman (2011). Emergence, singularities, and symmetry breaking. *Foundations of Physics*, 41 (6): 1031–1050.

Robert W. Batterman (2017). Philosophical implications of Kadanoff's work on the renormalization group. *Journal of Statistical Physics*, 167 (3–4): 559–574.

Robert W. Batterman (2018). Autonomy of theories: An explanatory problem. *Nous*, 52 (4): 858–873.

Robert W. Batterman (2019). Universality and RG explanations. *Perspectives on Science*, 27 (1): 26–47.

Robert W. Batterman and Sara Green (2020). Steel and bone: Mesoscale modeling and middle-out strategies in physics and biology. *Synthese*, DOI: 10.1007/s11229-020-02769-y.

Robert W. Batterman and Collin C. Rice (2014). Minimal model explanations. *Philosophy of Science*, 81 (3): 349–376.

Michael V. Berry (1987). The Bakerian Lecture: Quantum chaology. In M. V. Berry, I. C. Percival, and N. O. Weiss, editors, *Dynamical Chaos*, volume 186, pp. 183–198. Royal Society of London, Princeton University Press.

Charlotte Bigg (2008). Evident atoms: Visuality in Jean Perrin's brownian motion research. *Studies in History and Philosophy of Science*, 39: 312–322.

Cliff P. Burgess (2004). Quantum gravity in everyday life: General relativity as an effective field theory. *Living Reviews in Relativity*, 7 (1): 5. doi: 10.12942/lrr-2004-5. URL https://doi.org/10.12942/lrr-2004-5.

Jeremy Butterfield (2011a). Less is different: Emergence and reduction reconciled. *Foundations of Physics*, 41 (6): 1065–1135.

Jeremy Butterfield (2011b). Emergence, reduction and supervenience: A varied landscape. *Foundations of Physics*, 41: 920–959.

Carlo Cercignani (1988). *The Boltzmann Equation and Its Applications*. Applied Mathematical Sciences. Springer-Verlag.

Sergio Chibbaro, Lamberto Rondoni, and Angelo Vuilpiani (2014). *Reduction, Emergence and Levels of Reality: The Importance of Being Borderline.* Springer, Switzerland.

Richard M. Christensen (2005). *Mechanics of Composite Materials.* Dover, Mineola, New York.

Luca Cristofolini, Fulvia Taddei, Massimiliano Baleani, Fabio Baruffaldi, Susanna Stea, and Marco Viceconti (2008). Multiscale investigation of the functional properties of the human femur. *Philosophical Transactions of the Royal Society A: Mathematical, Physical and Engineering Sciences,* 366 (1879): 3319–3341.

D. D'Humières and P. Lallemand (1986). Lattice gas automata for fluid mechanics. *Physica A,* 140: 326–335.

Foad Dizadji-Bahamani, Roman Frigg, and Stephan Hartmann (2010). Who's afraid of Nagelian reduction? *Erkenntniss,* 73: 393–412.

John F. Donoghue (2012). The effective field theory treatment of quantum gravity. *AIP Conference Proceedings,* 1483 (1): 73–94. doi: 10.1063/1.4756964. URL https://aip.scitation.org/doi/abs/10.1063/1.4756964.

John Earman and Miklós Rédei (1996). Why ergodic theory does not explain the success of equilbrium statistical mechanics. *The British Journal for the Philosophy of Science,* 47: 63–78.

Peter Eastman. Introduction to statistical mechanics. https://peastman.github.io/statmech/, last accessed February 13, 2020.

Albert Einstein (1956). *Investigations on the Theory of the Brownian Movement.* Dover Publications.

Michael E. Fisher (1988). Condensed matter physics: Does quantum mechanics matter? In Herman Feshbach, Tetsuo Matsui, and Alexandra Oleson, editors, *Niels Bohr: Physics and the World,* volume Proceedings of the Niels Bohr Centennial Symposium, pp. 65–115, Chur, Switzerland, 1988. Harwood Academic Publishers GmbH.

Michael E. Fisher (1998). Renormalization group theory: Its basis and formulation in statistical physics. *Reviews of Modern Physics*, 70 (2): 653–681.

Jerry Fodor (1974). Special sciences, or the disunity of sciences as a working hypothesis. *Synthese*, 28: 97–115.

Jerry Fodor (1997). Special sciences: Still autonomous after all these years. *Philosophical Perspectives*, 11: 149–163.

Dieter Forster (1990). *Hydrodynamic Fluctuations, Broken Symmetry, and Correlation Functions.* Advanced Book Classics. Perseus Books.

Nigel Goldenfeld and Leo P. Kadanoff (1999). Simple lessons from complexity. *Science*, 284: 87–89.

Nelson Goodman (1983). *Fact, Fiction, and Forecast.* Harvard University Press, fourth edition.

E. A. Guggenheim (1945). The principle of corresponding states. *The Journal of Chemical Physics*, 13 (7): 253–261.

Hans Halvorson and Michael Mueger (2006). Algebraic quantum field theory.

Carl G. Hempel (1965). Empiricist criteria of cognitive significance: Problems and changes. In *Aspects of Scientific Explanation and Other Essays in the Philosophy of Science*, pp. 101–122. The Free Press.

David Hilbert (1902). Mathematical problems. *Bulletin of the American Mathematical Society*, 8: 437–479.

Christopher Hitchcock (2012). Events and times: A case study in means-ends metaphysics. *Philosophical Studies*, 160: 79–96.

David Jeffrey (1977). The physical significance of non-convergent integrals in expressions for effective transport properties. In J.W. Provan, editor, *Continuum Models of Discrete Systems*, pp. 653–673. University of Waterloo Press.

Leo P. Kadanoff (2000). *Statistical Physics: Statics, Dynamics, and Renormalization*. World Scientific, Singapore.

Leo P. Kadanoff (2013). Theories of matter: Infinities and renormalization. In Robert W. Batterman, editor, *The Oxford Handbook of Philosophy of Physics*, chapter 4, pp. 141–188. Oxford University Press.

Leo P. Kadanoff and Paul C. Martin (1963). Hydrodynamic equations and correlation functions. *Annals of Physics*, 24: 419–469.

Leo P. Kadanoff, Guy R. McNamara, and Gianluigi Zanetti (1989). From automata to fluid flow: Comparisons of simulation and theory. *Physical Review A*, 40 (8): 4527–4541.

Jaegwon Kim (1992). Multiple realization and the metaphysics of reduction. *Philosophy and Phenomenological Research*, 52 (1): 1–26.

Rami Koskinen (2019). Multiple realizability and biological modality. *Philosophy of Science*, 86 (5): 1123–1133.

Ryogo Kubo (1966). The fluctuation dissipation theorem. *Reports on Progress in Physics*, 29: 255–284.

Ryogo Kubo (1986). Brownian motion and nonequilibrium statistical mechanics. *Science*, 233 (4761): 330–334.

Marc Lange (2015). On "minimal model explanations": A reply to Batterman and Rice. *Philosophy of Science*, 82: 292–305.

David B. Malament and Sandy L. Zabell (1980). Why Gibbs phase averages work—the role of ergodic theory. *Philosophy of Science*, 47: 339–349.

N. Marcuvitz and J. Schwinger (1951). On the representation of the electric and magnetic fields produced by currents and discontinuities in wave guides. i. *Journal of Applied Physics*, 22 (6): 806–819.

Paul C. Martin and Julian Schwinger (1959). Theory of many-particle systems I. *Physical Review*, 115 (6): 1342–1373.

Calvin C. Moore (2015). Ergodic theorem, ergodic theory, and statistical mechanics. *Proceedings of the National Academy of Sciences*, 112 (7): 1907–1911.

Ernest Nagel (1961). *The Structure of Science: Problems in the Logic of Scientific Explanation*. Harcourt, Brace, & World.

Sia Nemat-Nasser and Muneo Hori (1999). *Micromechanics: Overall Properties of Heterogeneous Materials*. Elsevier, North-Holland, Amsterdam, second edition.

Denis Noble (2006). *The Music of Life: Biology Beyond Genes*. Oxford University Press.

Denis Noble (2011). A theory of biological relativity: No privileged level of causation. *Interface Focus*, 2 (1): 55–64.

Abraham Pais (1982). *Subtle is the Lord ... : The Science and Life of Albert Einstein*. Oxford University Press, Oxford.

David Papineau (1993). *Philosophical Naturalism*. Blackwell.

Thomas W. Polger and Lawrence A. Shapiro (2016). *The Multiple Realization Book*. Oxford University Press.

Hillary Putnam (1975). Philosophy and our mental life. In *Mind, Language and Reality: Philosophical Papers*, volume 2, pp. 291–303. Cambridge University Press, Cambridge.

Alexander Reutlinger, Dominik Hangleiter, and Stephan Hartmann (2018). Understanding (with) toy models. *The British Journal for the Philosophy of Science*, 69: 1069–1099.

Robert O. Ritchie, Markus J. Buehler, and Paul Hansma (2009). Plasticity and toughness in bone. *Physics Today*, 62 (6): 41–47.

Fereshteh A Sabet, Ahmad Raeisi Najafi, Elham Hamed, and Iwona Jasiuk (2016). Modelling of bone fracture and strength at different length scales: a review. *Interface focus*, 6 (1): 20150055.

Kenneth Schaffner (2013). Ernest Nagel and reduction. *The Journal of Philosophy*, 109 (534–565).

Julian Schwinger (1963). Fields and particles. Lecture at the Belfer Graduate School of Science, Yeshiva University. Box 15, Folder 39: Julian Seymour Schwinger Papers (Collecrtion 371). Department of Special Collections, Charles E. Young Research Library, University of California, Los Angeles.

Julian Schwinger (1969). Julian Schwinger's engineering approach to particle theory. *Scientific Research*, 4 (17): 19–24.

Jonathan Selinger (2016). *Introduction to the Theory of Soft Matter: From Ideal Gases to Liquid Crystals*. Springer.

James P. Sethna (2006). *Statistical Mechanics: Entropy, Order Parameters, and Complexity*. Oxford Master Series in Statistical, Computational, and Theoretical Physics. Oxford University Press.

Theodore Sider (2011). *Writing the Book of the World*. Oxford University Press.

Lawrence Sklar (1993). *Physics and Chance: Philosophical Issues in the Foundations of Statstical Mechanics*. Cambridge University Press, Cambridge.

Elliott Sober (1999). The multiple realizability argument against reductionism. *Philosophy of Science*, 66: 542–564.

Tuomas E. Tahko (2018). Fundamentality. In Edward N. Zalta, editor, *The Stanford Encyclopedia of Philosophy*. Metaphysics Research Lab, Stanford University, fall 2018 edition.

Andrei Tokmakoff. Time-dependent quantum mechanics and spectroscopy. https://tdqms.uchicago.edu last accessed March 2, 2020.

Salvatore Torquato (2002). *Random Heterogeneous Materials: Microstructure and Macroscopic Properties*. Springer, New York.

Clifford Truesdell (1984). *An Idiot's Fugitive Essays on Science: Methods, Criticism, Training, Circumstances*. Springer-Verlag, New York.

Clifford Truesdell and Walter Noll (1992). *The Non-Linear Field Theories of Mechanics*. Springer-Verlag, 2nd edition.

Jos Uffink (2017). Boltzmann's work in statistical physics. In Edward N. Zalta, editor, *The Stanford Encyclopedia of Philosophy*. Metaphysics Research Lab, Stanford University, spring 2017 edition.

Charlotte Werndl and Roman Frigg (2015a). Reconceptualising equilibrium in Boltzmannian statistical mechanc and characterising its existence. *Studies in the History and Philosophy of Modern Physics*, 49: 19–31.

Charlotte Werndl and Roman Frigg (2015b). Rethinking Boltzmann equilibrium. *Philosophy of Science*, 82 (5): 1224–1235.

Porter Williams (2019). Scientific realism made effective. *The British Journal for the Philosophy of Science*, 70: 209–237.

Mark Wilson (2018). *Physics Avoidance: And Other Essays in Conceptual Strategy*. Oxford University Press.

James Woodward (2005). *Making things happen: A theory of causal explanation.* Oxford University Press.

James Woodward (2016). The problem of variable choice. *Synthese*, 193: 1047–1072.

Index

Note: Page references followed by a "*t*" indicate table; "*f*" indicates figure.